IMAGES
of America

CORNWALL

On the Cover: Pictured around 1950, most of these miners are open-pit miners transitioning to underground work as the large open pit was closing. (Courtesy of the authors.)

IMAGES
of America

CORNWALL

Charles Boyer and Robert Donley

ARCADIA
PUBLISHING

ISBN 978-0-7385-7465-3

Published by Arcadia Publishing
Charleston, South Carolina

Printed in the United States of America

Library of Congress Control Number: 2010937246

For all general information, please contact Arcadia Publishing:
Telephone 843-853-2070
Fax 843-853-0044
E-mail sales@arcadiapublishing.com
For customer service and orders:
Toll-Free 1-888-313-2665

Visit us on the Internet at www.arcadiapublishing.com

This book is proudly dedicated to the honor of the men of steel who faithfully mined the iron ore from 1737 to 1973.

Contents

Acknowledgments		6
Introduction		7
Cornwall Time Line		9
1.	The Early Years	11
2.	Steam Trains Come to Cornwall	15
3.	Steam Shovels Come to Cornwall	33
4.	Bethlehem Steel and Cornwall	47
5.	Churches, Businesses, and the Community of Cornwall	77
6.	It Is All Underground Now	93

ACKNOWLEDGMENTS

This book is a general overview of the community of Cornwall. Much time could have been spent on the history of the families that became very wealthy because they owned the iron ore mines, the houses where the miners lived, and most of the land in what is now Cornwall prior to the Bethlehem Steel Corporation's purchase of the mine. We, however, chose to focus mostly on the people who worked in the mines and some businesses that supported the miners and the community.

We would like to thank the following for their help in developing the information for this book: the Cornwall Iron Furnace State Historic Site and the *Lebanon Daily News* for the use of information; Milan Lipensky, Charles Petry, Patricia Herschkowitz, and Rev. Gary Beard for the time they granted for interviews and the use of information from those interviews; and the Joseph P. Manypenny Memorial Library at Cornwall Manor, Daniel Chernich, Mr. and Mrs. Anthony Savant, Mr. and Mrs. Ray Dishong, Jean Kotkas, and Robert Boyer for the use of vintage pictures. Special thanks go to Douglas H. Haack, owner of Digital PhotoGenics, LLC, in Rancho Palos Verdes, California, for his talent and work in making an otherwise unusable image usable.

We would like to give a big thank-you to our wives, Jean and Lucinda, for their help, support, and encouragement in writing this book. We would like to thank our families for their support and valued comments along with the proofreading of this book. Throughout the book, the owner of an image has been given credit. Images without a credit are from the authors' private collections.

We would like to thank Arcadia Publishing for publishing this book. The workings in Cornwall have been closed since 1973. Fewer and fewer miners are still living. Without this book, their story might have been forgotten and lost.

INTRODUCTION

Welcome to Cornwall, Pennsylvania. Peter Grubb named Cornwall for the area in England of the same name where his father was born. In three now famous hills, Grubb discovered outcroppings of the iron ore deposit that would be mined well into the 20th century. His find in 1737 marked the start of the longest continuously operating mining operation in the world. Any contending mines would have to beat 236 years of continuous work. Shortly after his discovery, Grubb bought 300 acres of the surrounding land for 135 pounds from William Allen, Pennsylvania's chief justice at the time. In 1737, Grubb was granted a warrant for an additional 142.5 acres, and it was in this parcel that the three famous hills—Big Hill, Middle Hill, and Grassy Hill—sat. That year, Grubb also built two forges just south of the ore deposit on Hammer Creek. The forges began smelting ore from the new mine.

In 1742, Grubb built a charcoal furnace adjacent to the mine. The furnace could produce more iron than the forges, allowing more ore to be processed. Named the Cornwall Iron Furnace, it operated for 140 years, closing in 1883. That furnace remains intact and is maintained by the Pennsylvania Historical and Museum Commission, which offers guided tours.

The history of the community of Cornwall and the surrounding villages is so intertwined with the history of the mine that it is impossible to deal with them independently.

Cornwall was built as a result of mining, and villages were located close to the worksites. Minersvillage was built by the Cornwall Ore Bank Company between 1864 and 1900, and the houses were rented to the workers. The village of Anthracite was built by brothers Richard W. and William Coleman between 1851 and 1853, and the homes were rented to workers at the Anthracite Furnace. The villages of Burd Coleman and North Cornwall were built for and houses rented to workers at the Burd Coleman and North Cornwall Furnaces. These three furnaces were fueled by anthracite coal rather than the wood charcoal used in the original Cornwall Iron Furnace. Rexmont was established between 1880 and 1900 by workers at the mine who wanted to own their own homes rather than rent a house from the mining company. Rexmont was built on land cleared by the woodcutters and colliers who cut timber and made charcoal for the fuel at the Cornwall Iron Furnace.

Each village had a school for the miners' children or was near an existing school. Most villages had their own stores as well, typically run by enterprising folks and not the company. Originally, the village schools were independent of each other, but early in the 1900s, the smaller schoolhouses were closed, and one larger school was opened in the center of Cornwall. Churches, an important aspect to the community, were also centrally located.

In 1926, Bethlehem Steel Company became sole owner of the mines, land, mineral rights, and homes. The mines at Cornwall extracted iron ore until 1973. In 1972, Hurricane Agnes flooded the underground No. 3 Mine. Ore was still there to be mined in both underground mines for a few more years, but the expense of opening the mine again after the flood might have equaled or exceeded the value of the ore remaining. A small open-pit operation excavating the east-end ore body continued working until 1973 when the iron ore was depleted in that particular deposit.

The iron ore pellet plant that opened in 1962 to make pellets out of Cornwall iron ore continued operating after the flood for a few more years by processing foreign minerals rather than Cornwall ore. The pellets, efficient at heating by allowing air to flow around their round shapes, were used in the blast furnaces at other Bethlehem Steel locations.

In its latter years, Bethlehem Steel employed approximately 900 people at Cornwall, including underground miners working the No. 3 and No. 4 Mines, open-pit miners at the east-end ore body, pellet plant employees, and administrative employees. Bethlehem Steel closed mining forever at Cornwall in 1973, thus ending 236 continuous years of mining in the area. The remaining company landholdings were sold to developers who built other villages and improvements. There are now 16 separate villages and other developed areas in Cornwall, which covers 9.9 square miles.

Many families in Cornwall can count generations of miners in their histories. It is not unusual to find three or four generations of miners in one family.

Cornwall had three railroads. The Cornwall Ore Bank Railroad was used strictly for hauling out ore from the open pit, and the Cornwall Railroad and Cornwall & Lebanon Railroads carried ore to blast furnaces, passengers, and freight. The Cornwall Railroad transported five million riders during its operating years. The passengers were able to travel to local parks, such as Penryn Park and Mount Gretna, or to Lebanon, Pennsylvania, for connection to other cities.

Cornwall Time Line

1730: Peter Grubb begins prospecting for iron ore in an area that was part of northern Lancaster County.

1732: Grubb discovers three hills (Big Hill, Middle Hill, and Grassy Hill) containing magnetite iron ore in an area that is now part of southern Lebanon County.

1734: Grubb purchases 300 acres of land from William Allen for 135 pounds.

1737: Grubb is granted a warrant for an additional 142.5 acres of land. The three hills containing magnetite ore are located on this land.

1737: Grubb builds two forges on Hammer Creek. These forges are used to smelt ore extracted from the mine at Cornwall.

1742: Grubb builds a charcoal furnace adjacent to the iron ore deposit and names it Cornwall Iron Furnace.

1754: Grubb dies and ownership of the Cornwall Iron Furnace and the mine passes to his sons Curttis and Peter.

1777: German mercenary (Hessian) soldiers captured by George Washington's army at the Battle of Trenton on December 26, 1776, are used for labor at the Cornwall Iron Furnace and are paid 60¢ a day.

1783: Curttis Grubb grants eldest son, called Peter Jr., a one-sixth interest in the ore deposit.

1786: Peter Grubb Jr. grants his one-sixth interest to Robert Coleman. Grubb reserved the right for himself and his heirs to obtain a sufficient quantity of ore to supply one furnace. This right became known as the Robesonia Iron Company.

1798: Through acquisitions from the Grubb Estate, Robert Coleman acquires five-sixths ownership of the ore deposits. The remaining one-sixth belonged to Henry Bates Grubb (son of Peter Grubb Jr.).

1825: Robert Coleman dies. His five-sixths interest passes to sons William, James, Edward, and Thomas Burd Coleman.

1856: The Cornwall Iron Furnace's melt furnace is rebuilt. The rebuilt furnace is 32 feet high with a seven-foot bosh and a three-foot crucible.

1864: The Cornwall Ore Bank Company is formed with J. Taylor Boyd in charge of all mining operations except the portion mined by the Robesonia Iron Company. The formation of the Cornwall Ore Bank Company was the result of a lawsuit that started in 1851 and concluded in

1869. There were 96 shares of ownership in the Cornwall Ore Bank Company. All share owners would, for the first time, be able to receive an accurate accounting of the amount of ore mined. There were approximately six furnaces operating in the area at that time.

1883: After more than 140 years of operation, the last cast is made at the original Cornwall Iron Furnace, built by Peter Grubb.

1905: The Pennsylvania Steel Company, which is part owner of the Cornwall Ore Bank Company, builds a concentrator plant in Lebanon, Pennsylvania, to process the ore.

1906: Nodulizing of iron concentrates begins at the Lebanon Concentrator Plant and continues until 1920.

1910: Mechanization begins. Steam shovels are now used for loading. The old wheelbarrow methods are abandoned. The Cornwall Ore Bank Company constructs a crushing and screening plant. A 45-degree hoisting incline is built on the footwall side of the pit, thereby reducing the long rail haul.

1912: The flotation process of chalcopyrite is initiated at the Lebanon Concentrator Plant.

1916: Bethlehem Steel Company purchases the Pennsylvania Steel Company. That purchase includes the Lebanon Concentrator Plant and part ownership in the Cornwall Ore Bank Company. The concentrator plant is modernized with the addition of sintering pans.

1919: Robert Coleman's great-granddaughters, Isabelle Coleman Freeman and Margaret Coleman Buckingham, sell their interests in the mines to Bethlehem Steel Company.

1919: A second ore body (No. 4 Mine) is discovered.

1921: A three-compartment, 29-degree shaft is sunk into the footwall of the west end of the open pit used for mining the lower portion of the ore body. It is known as No. 3 Mine.

1922: The sintering operation is expanded at the Lebanon Concentrator Plant, and production is increased to 2,400 tons a day.

1926: Bethlehem Steel Company acquires the Robesonia Iron Company and becomes the sole owner of the Cornwall mines.

1927: Development of No. 4 Mine begins with the sinking of two inclined, three-compartment shafts in the footwall beneath the ore body.

1927: Cornwall becomes a borough.

1931: Work at No. 4 Mine is stopped because of the Depression. The mine is allowed to fill with water.

1931: Margaret Coleman Buckingham donates the Cornwall Iron Furnace, built by Peter Grubb, to the Pennsylvania Historical and Museum Commission. A $35,000 endowment is also given to the commission.

1937: Work is again started at No. 4 Mine.

1940: Underground work is stopped, and No. 3 Mine is closed because of open-pit workings.

1945: Development of No. 4 Mine reaches the 700-foot level, and it becomes necessary to deepen the shafts, which are extended to a final depth of 1,225 feet. This work was completed in 1948.

1950: The first commercial pelletizing plant in the United States is completed at the Lebanon Concentrator Plant.

1953: After 216 years of operation, the open pit is closed. The No. 3 Mine is reopened to extract the remaining ore body not accessible using open-pit methods.

1953: Two additional pellet furnaces are installed at the Lebanon Concentrator Plant.

1962: Bethlehem Steel Company constructs and opens a modern mill and pelletizing plant in Cornwall near No. 4 Mine.

1962: The Lebanon Concentrator Plant is closed after 57 years of operation.

1965: Work is begun to extend the pit eastward so that the east-end ore body can be mined using the open-pit method.

1972: Underground mining is stopped because of flooding from Hurricane Agnes and the loss of electricity.

1973: Mining operations cease. The open-pit mine for the east-end ore body has exhausted its deposits, and the remaining underground ore in the flooded mine is deemed too costly to remove. All mining in Cornwall is closed forever.

One

The Early Years

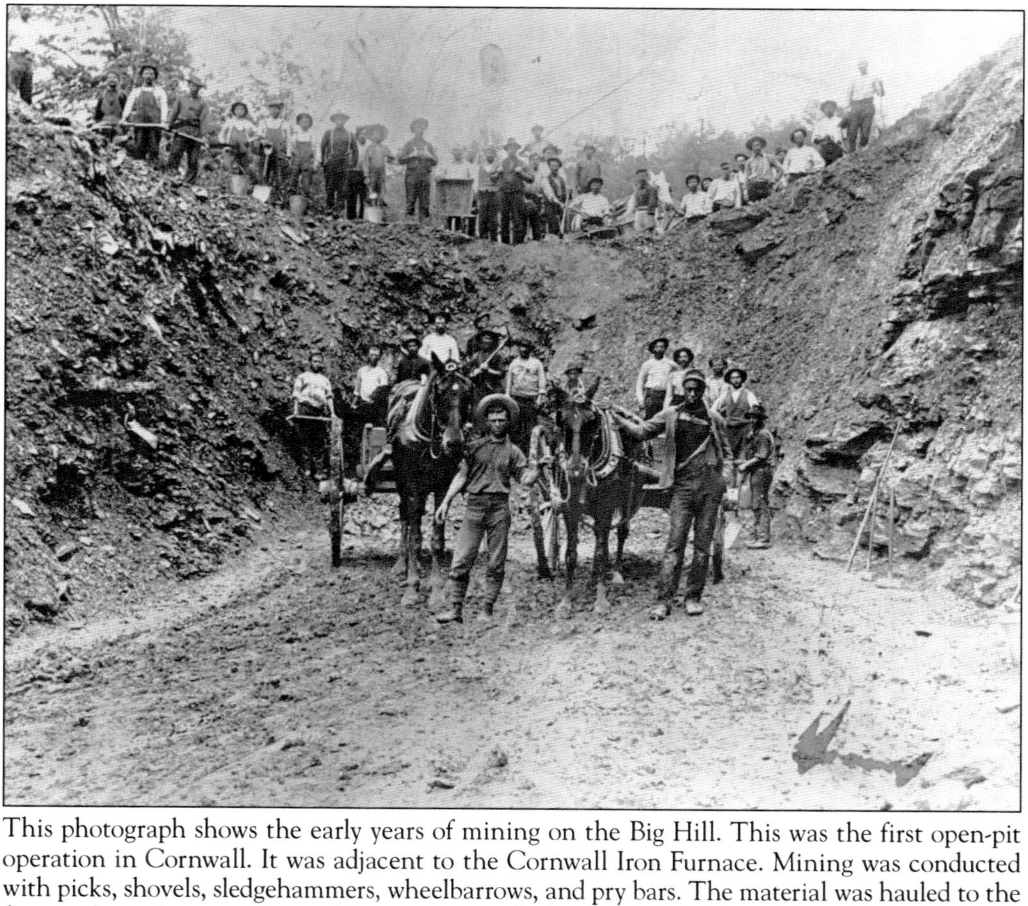

This photograph shows the early years of mining on the Big Hill. This was the first open-pit operation in Cornwall. It was adjacent to the Cornwall Iron Furnace. Mining was conducted with picks, shovels, sledgehammers, wheelbarrows, and pry bars. The material was hauled to the furnace by mule carts. Ore and limestone were both taken to the furnace. Limestone was used in the furnace to remove impurities from the molten iron.

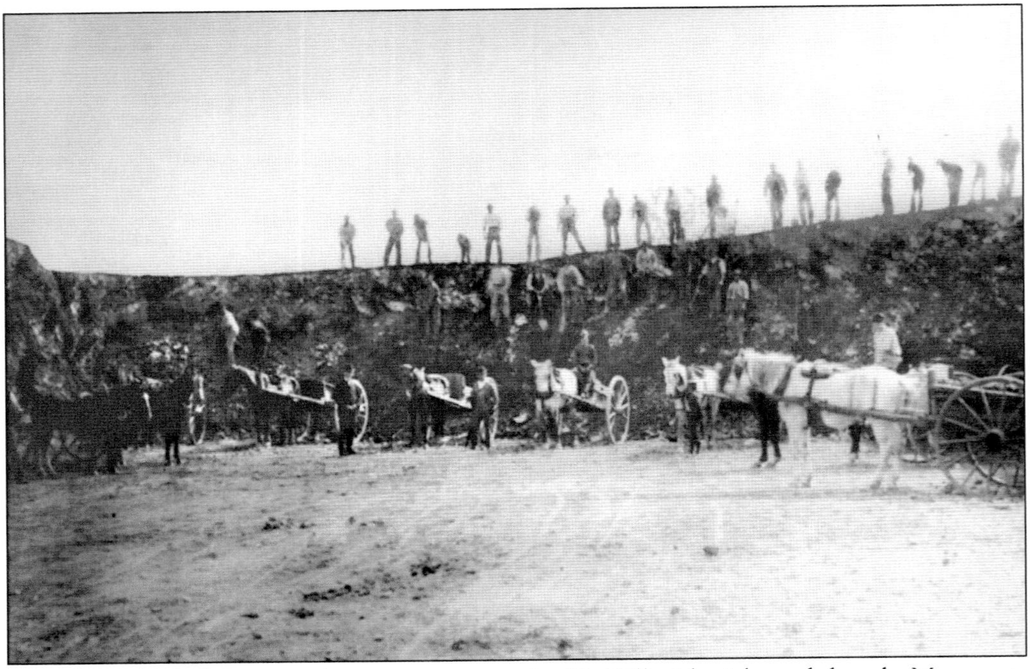

This photograph shows a work crew mining on the Big Hill with picks and shovels. More carts with teamsters are ready to haul material to the furnace to be melted. In those days, the iron was used to cast items such as stoves and frying pans, or it was made into a raw product known as pig iron, which was used by blacksmiths who made horseshoes and other household items.

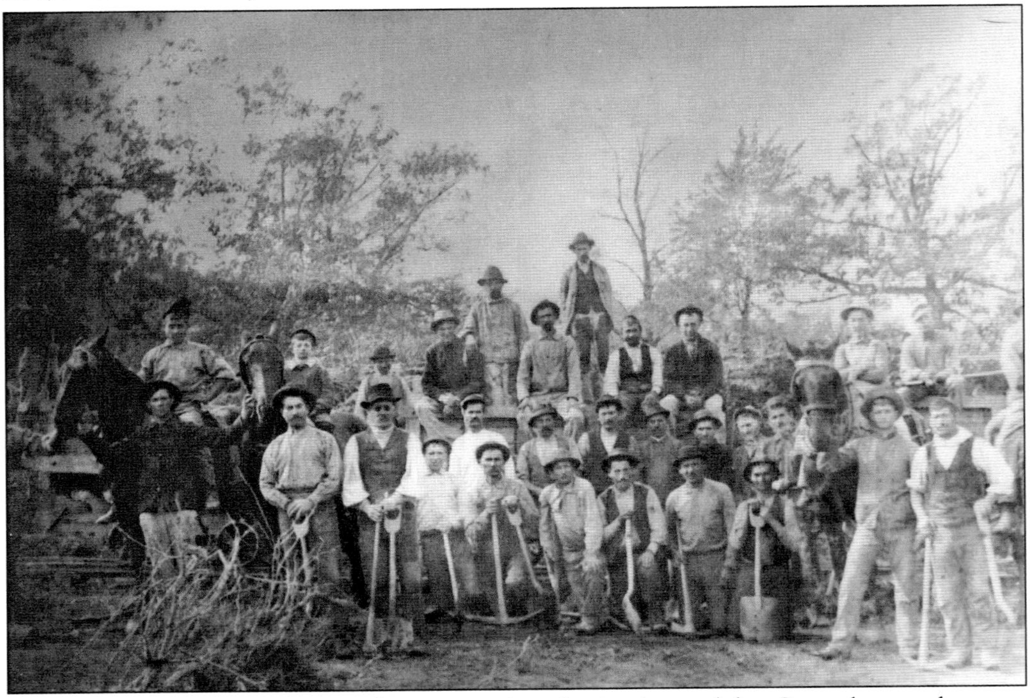

This photograph shows a group of miners ready to begin the workday. Several young boys are riding or tending horses as well. Young people also appear to be standing or sitting on the wagon. Mining was difficult work for men, and it must have been even harder labor for young boys.

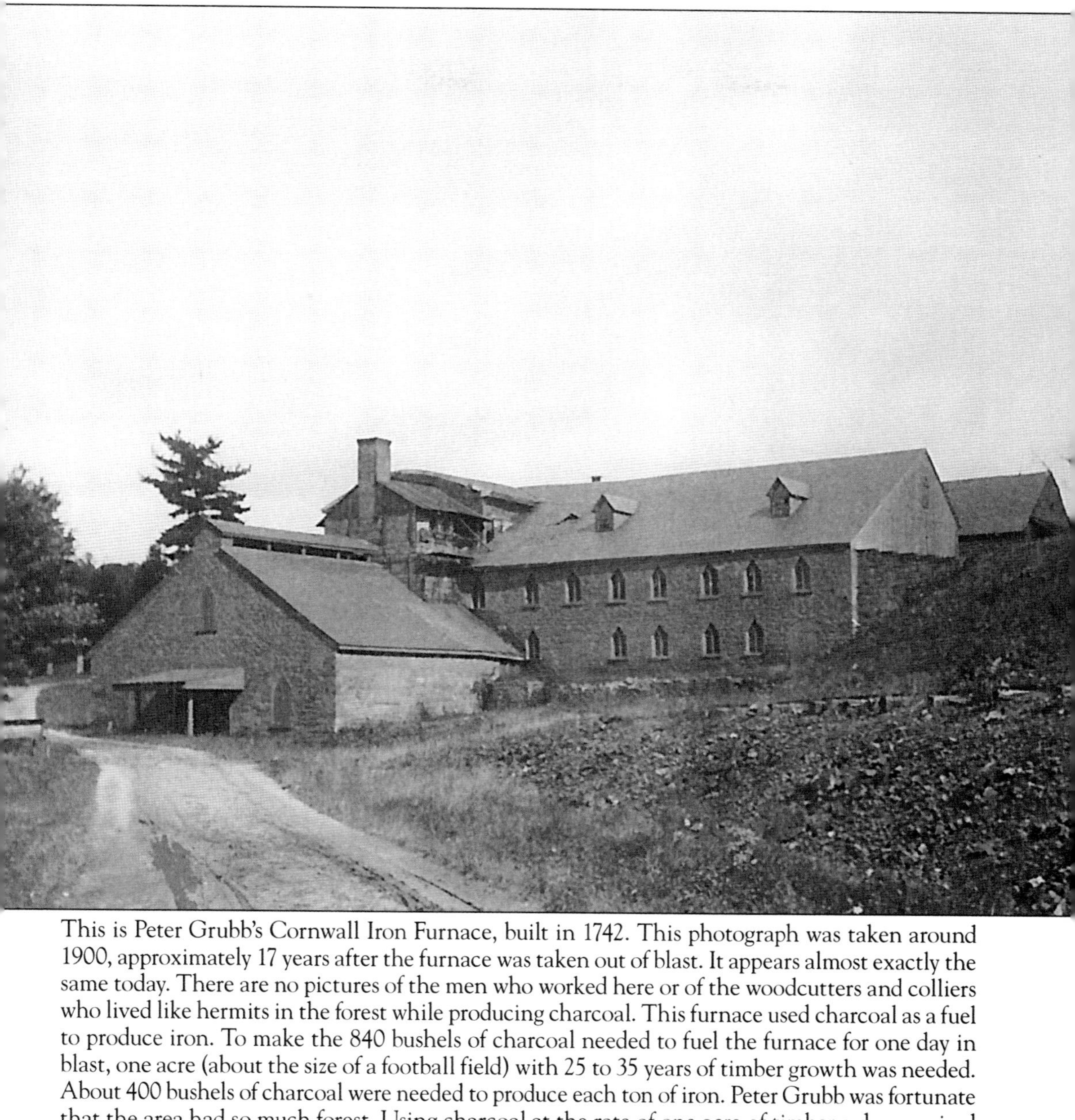

This is Peter Grubb's Cornwall Iron Furnace, built in 1742. This photograph was taken around 1900, approximately 17 years after the furnace was taken out of blast. It appears almost exactly the same today. There are no pictures of the men who worked here or of the woodcutters and colliers who lived like hermits in the forest while producing charcoal. This furnace used charcoal as a fuel to produce iron. To make the 840 bushels of charcoal needed to fuel the furnace for one day in blast, one acre (about the size of a football field) with 25 to 35 years of timber growth was needed. About 400 bushels of charcoal were needed to produce each ton of iron. Peter Grubb was fortunate that the area had so much forest. Using charcoal at the rate of one acre of timber a day required a lot of trees to operate the furnace for 140 years. The 8 to 12 men who worked each shift tapped the furnace noon and midnight. (Courtesy of the Cornwall Manor Manypenny Library.)

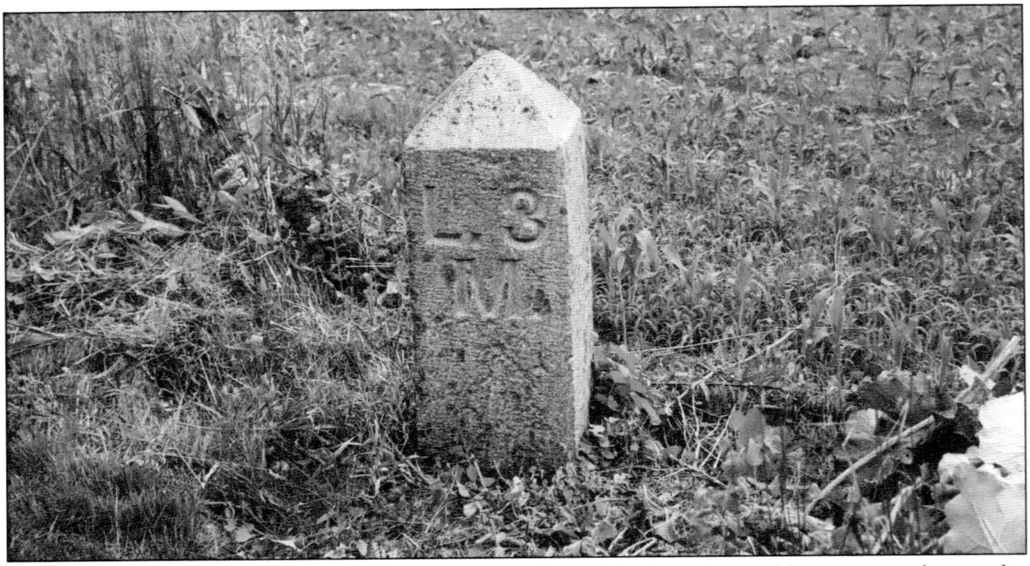

Iron products from the furnace and iron ore from the mine were shipped by wagon either to the loading docks of the Union Canal in Lebanon or to other furnaces in the area. The wagons had six-inch-wide wheels, but even with the extra width, they would leave large ruts in the road during wet weather. On April 11, 1848, the Pennsylvania State Legislature authorized the construction of the North and South Lebanon Turnpike. The North and South Lebanon Turnpike Company incorporated and built a plank road from the Union Canal in Lebanon to the Cornwall Ore Bank mine. Planks were cut, laid perpendicular to the road, and caulked with clay. A tollhouse was built to collect tolls so that the company could make a profit and maintain the road. Today, the highway known as the Cornwall Pike follows the route of the old wooden road. These photographs show one of the markers still in place. On one side, the number of miles to Lebanon appears (above), and on the other side, the number of miles to Cornwall is shown (below).

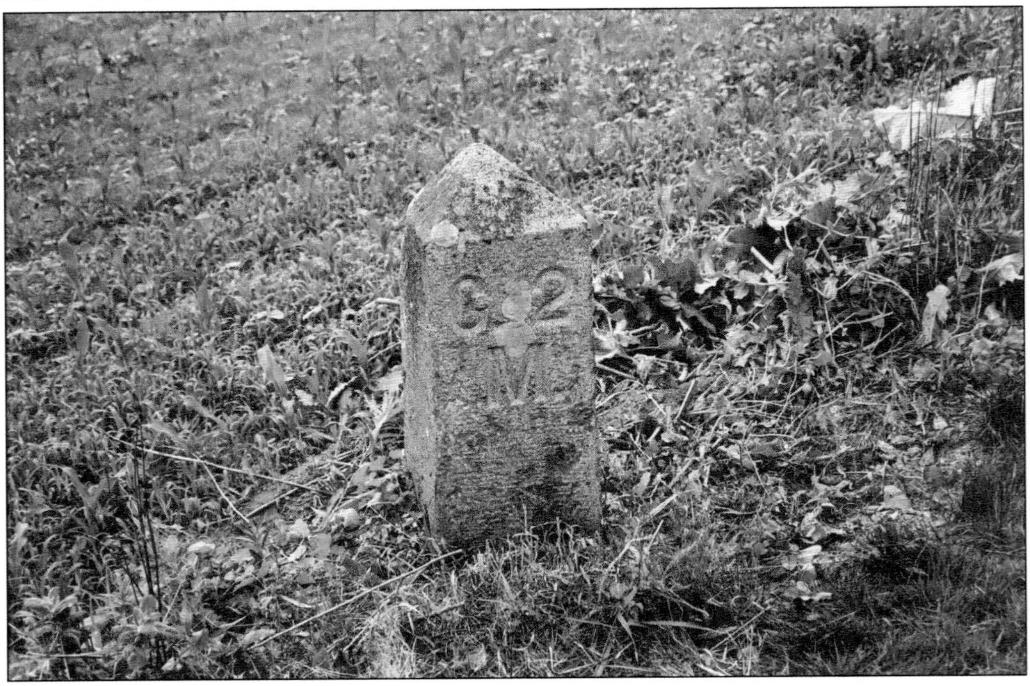

Two

Steam Trains Come to Cornwall

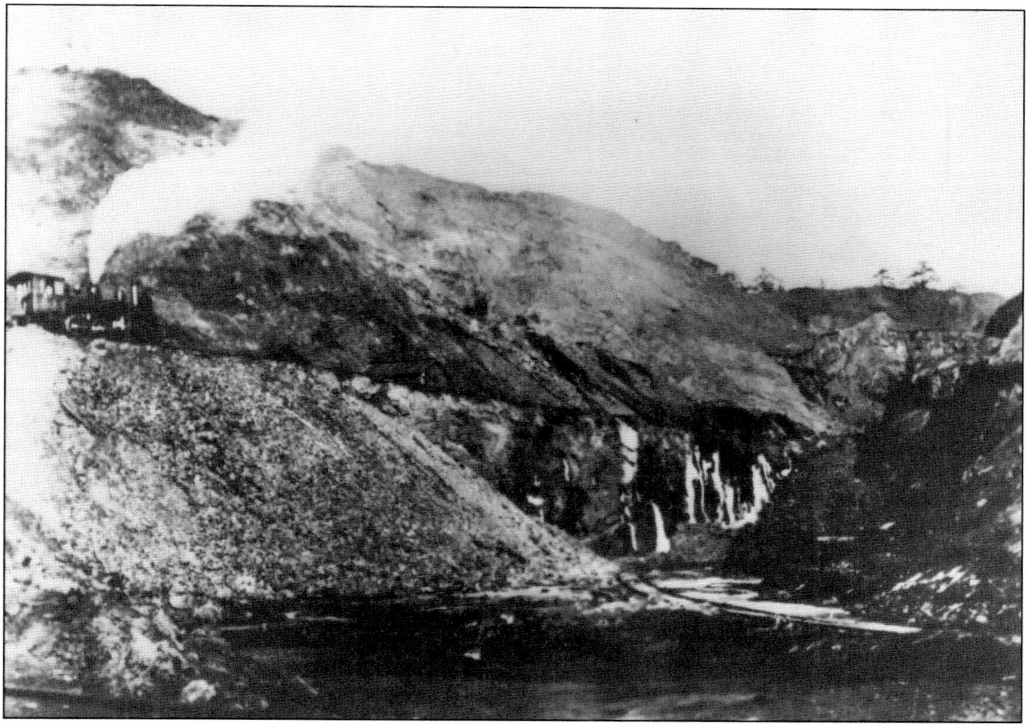

This eastward view shows Big Hill. Train tracks are visible in the distance at the center of the image. Steam trains replaced carts and mules to haul ore to the furnace during the mid-1800s. The picture also shows a train operating on an upper track of the system. Productivity increased dramatically with use of steam engines. The mine produced 776,000 tons of ore from 1742 to 1848; but from 1848 to 1864, more than 1.5 million tons of ore were mined. The owners at the time were becoming very wealthy indeed with improvements in output. The miners in the early years had needed to remove all the waste rock to get to the iron ore deposit. No mention is made of how many tons of such rock had to be taken away.

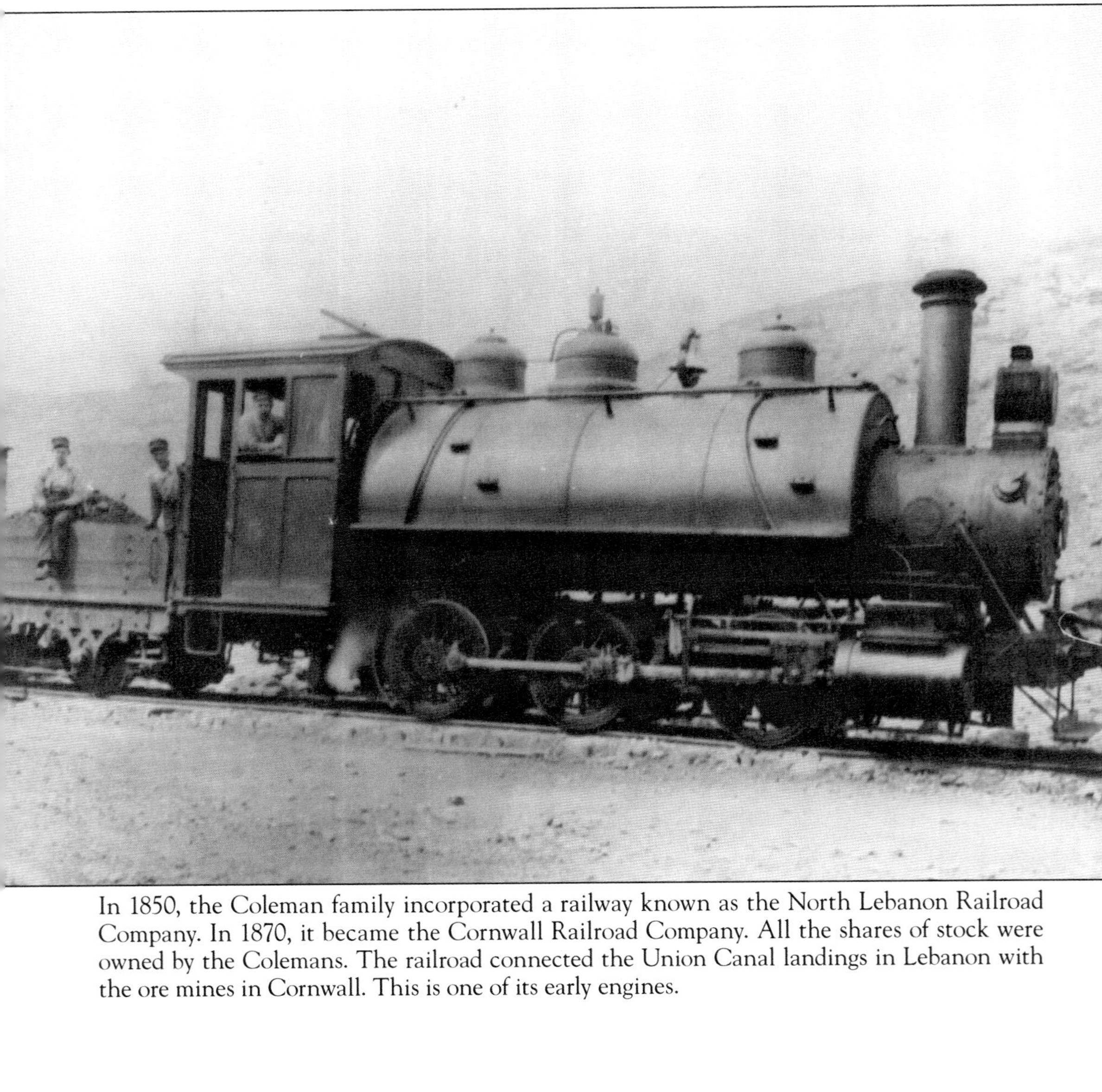

In 1850, the Coleman family incorporated a railway known as the North Lebanon Railroad Company. In 1870, it became the Cornwall Railroad Company. All the shares of stock were owned by the Colemans. The railroad connected the Union Canal landings in Lebanon with the ore mines in Cornwall. This is one of its early engines.

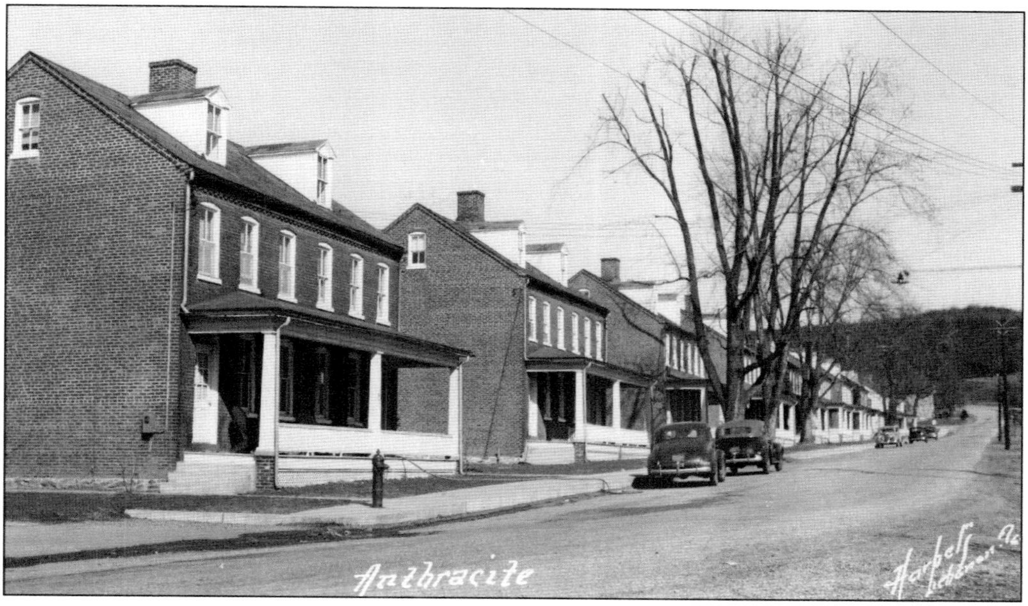

This is an early-1940s postcard of the village of Anthracite (above). Today, it is known as Goosetown. Brothers Robert W. and William Coleman built this village during the years 1851 to 1853 to house workers employed at the Anthracite Furnace. Part of the village of Anthracite includes a group of houses known as the Brick Row. These houses are pictured as they appear today (below).

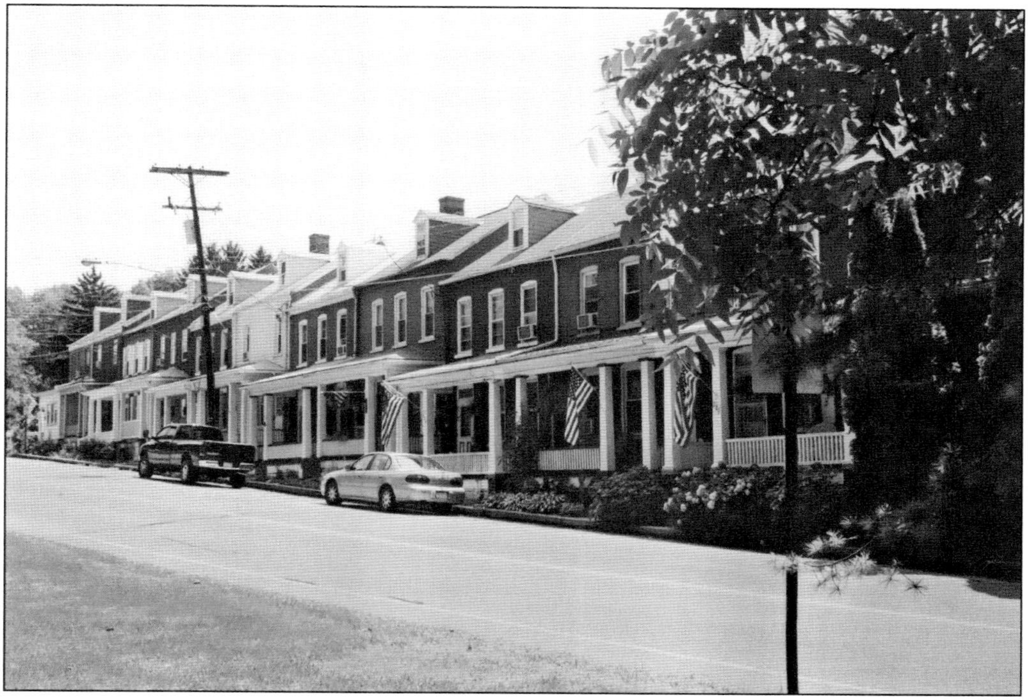

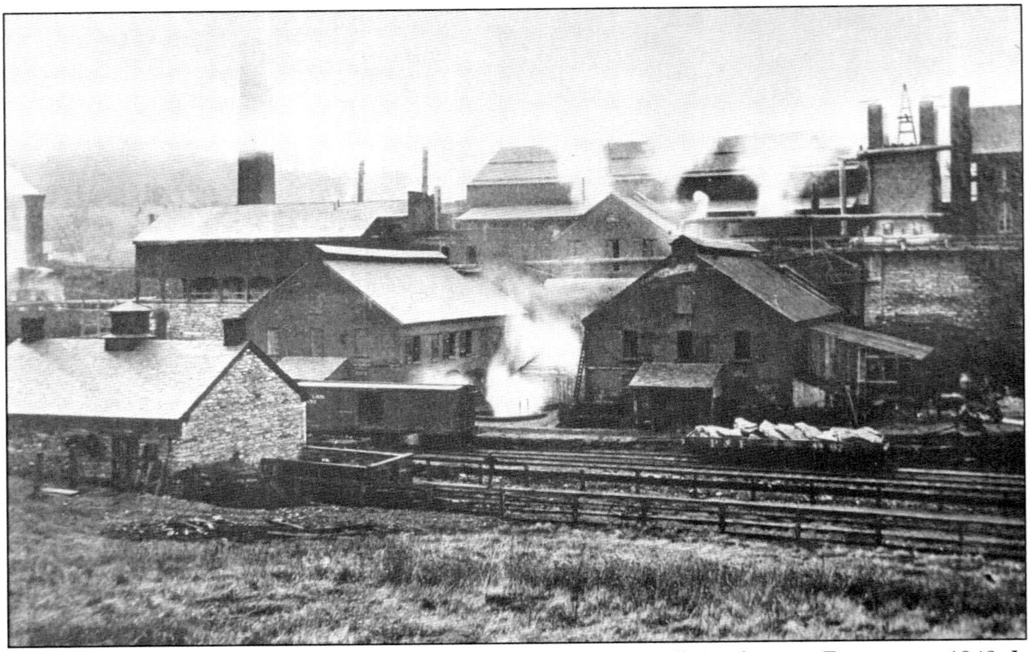

Robert W. and William Coleman began building the Cornwall Anthracite Furnace in 1849. It was the first of the modern furnaces that were designed to use anthracite coal as a fuel instead of the wood charcoal used in the first furnace. The furnace continually expanded until 1854 and was later purchased by the Lackawanna Iron and Steel Company of Scranton.

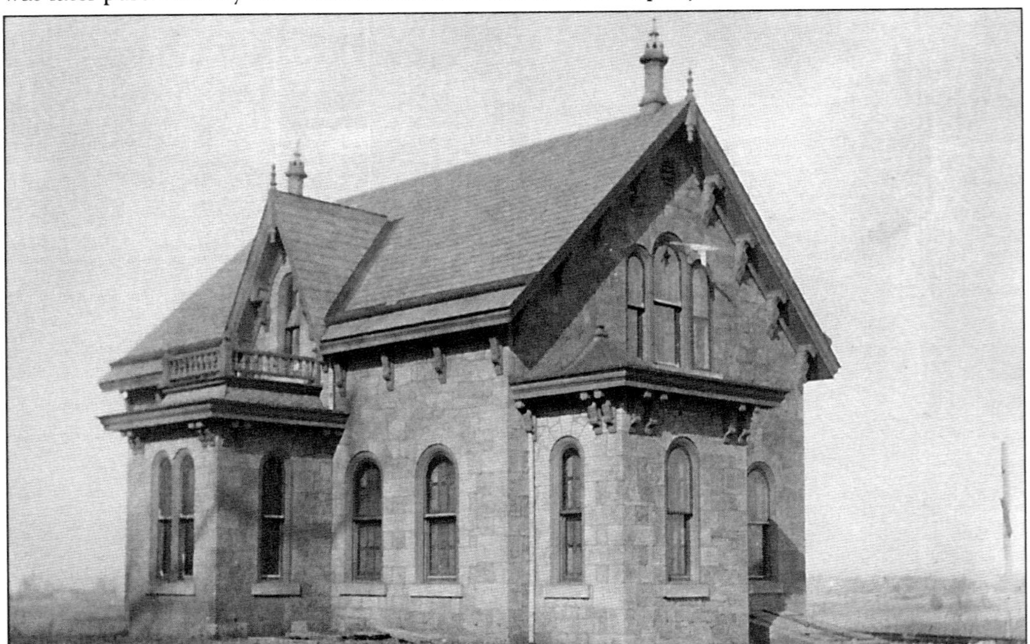

The Cornwall Anthracite Furnace was razed in 1899 after sitting idle for a number of years. The only things remaining are the village of Anthracite shown on the preceding page and the sandstone office in this photograph. The office was eventually purchased for conversion into a home and remains a private residence today. (Courtesy of Cornwall Manor Manypenny Library.)

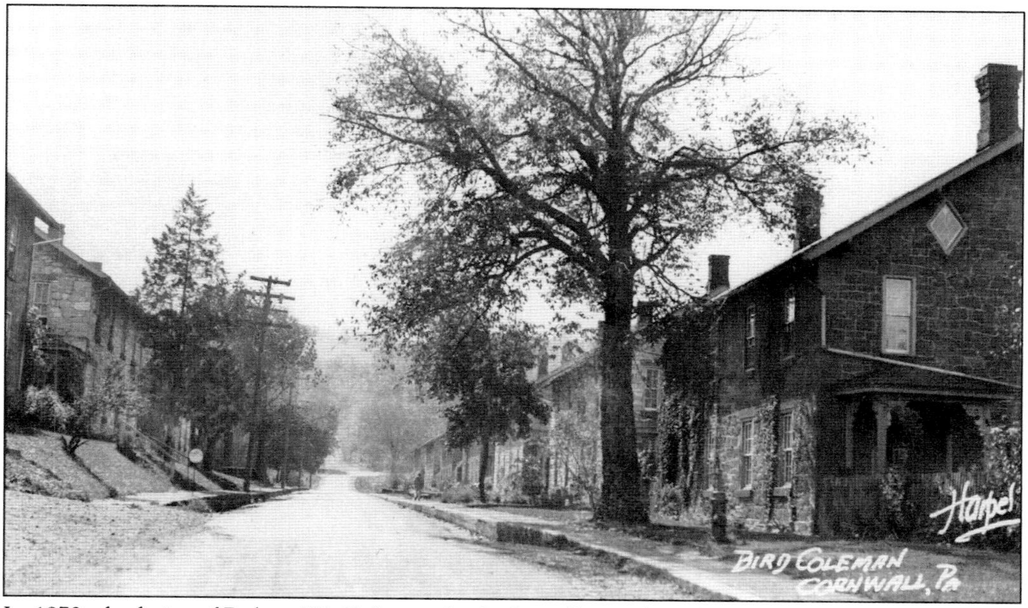

In 1872, the heirs of Robert W. Coleman built the village of Burd Coleman. Shown here in the early 1940s, the village still looks the same today.

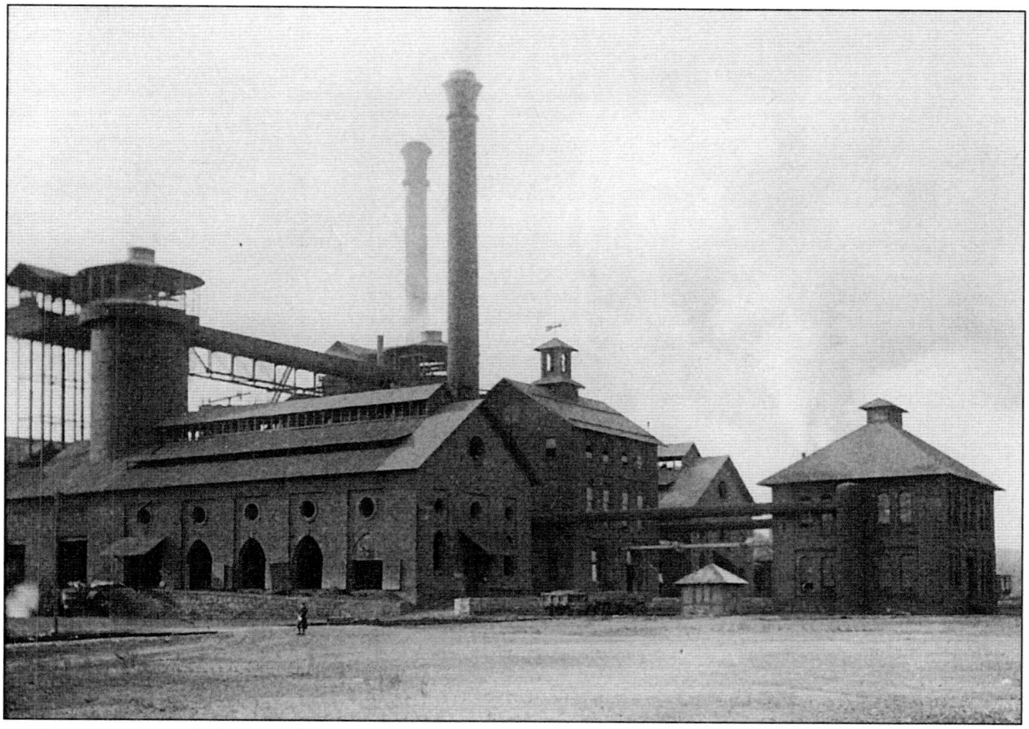

In the same year, the heirs of Robert W. Coleman built a new anthracite-fueled furnace for smelting ore. Smelting methods were improving, and the furnace was remodeled in 1880. In 1886, the owners incorporated under the name Cornwall Iron Company, Ltd. In 1887, the Lackawanna Iron and Steel Company took over management of the furnace by lease. It was sold to the Bethlehem Steel Company in 1918. The furnace was completely dismantled in 1922, and no trace of it remains. (Courtesy of Cornwall Manor Manypenny Library.)

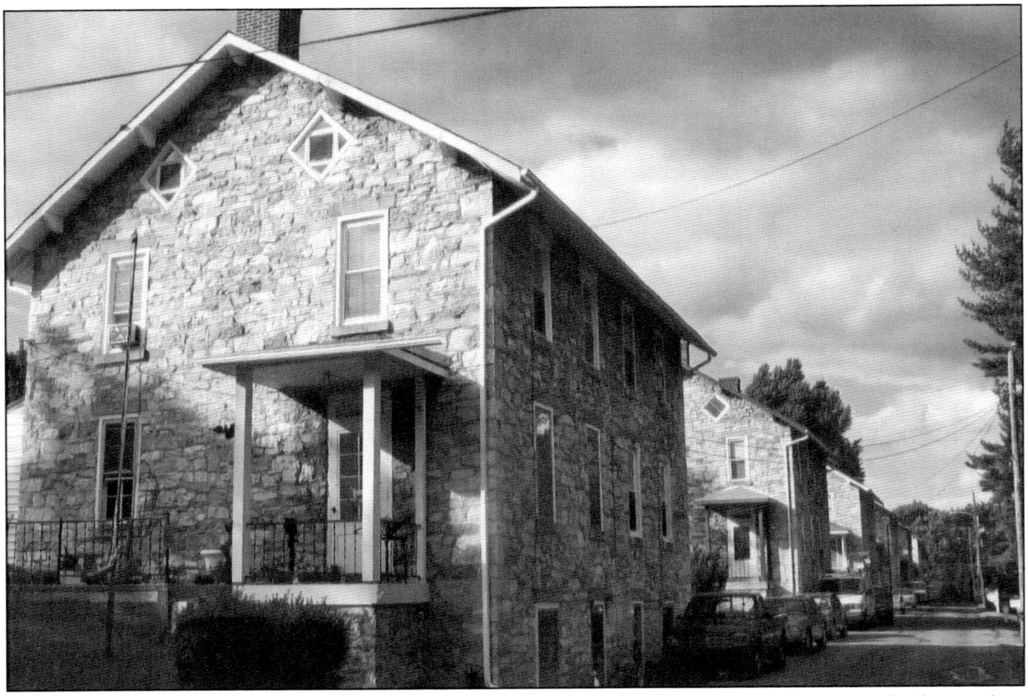

During 1872–1873, the village of North Cornwall was built. This recent photograph shows that it looks about the same today as when it was founded.

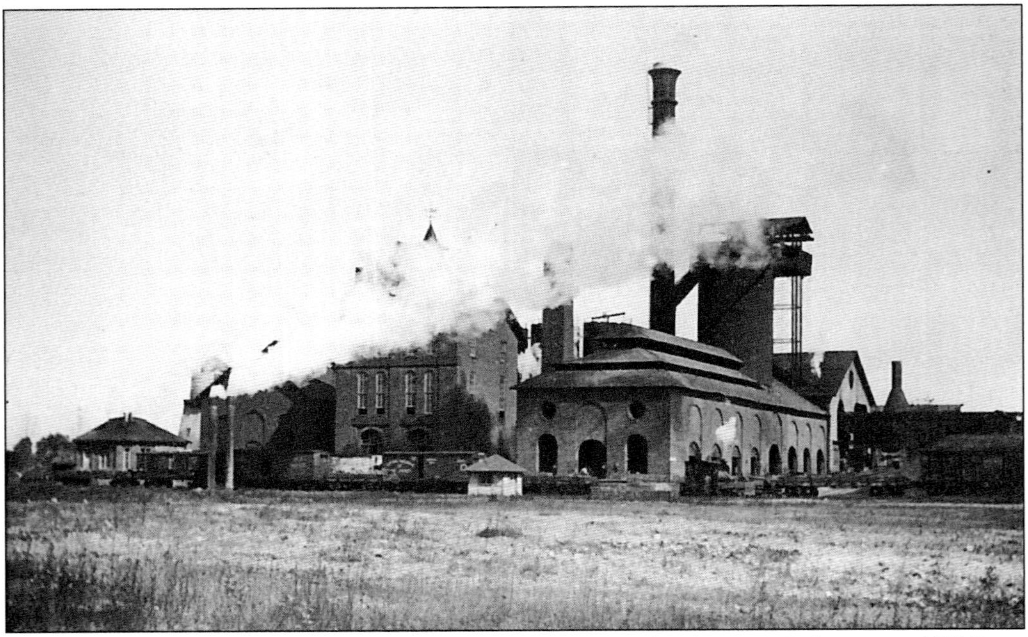

During the same time period, Margaret Freeman, a sister of Robert W. Coleman and William Coleman, built the modern, anthracite-fueled North Cornwall Furnace near the village of North Cornwall. Remodeled in 1887, the furnace was leased to the Lackawanna Iron and Steel Company in 1897. The Bethlehem Steel Company bought the facility in 1918 and tore it down in 1923. The only structure remaining is the office pictured on the next page, now a private residence. (Courtesy of Cornwall Manor Manypenny Library.)

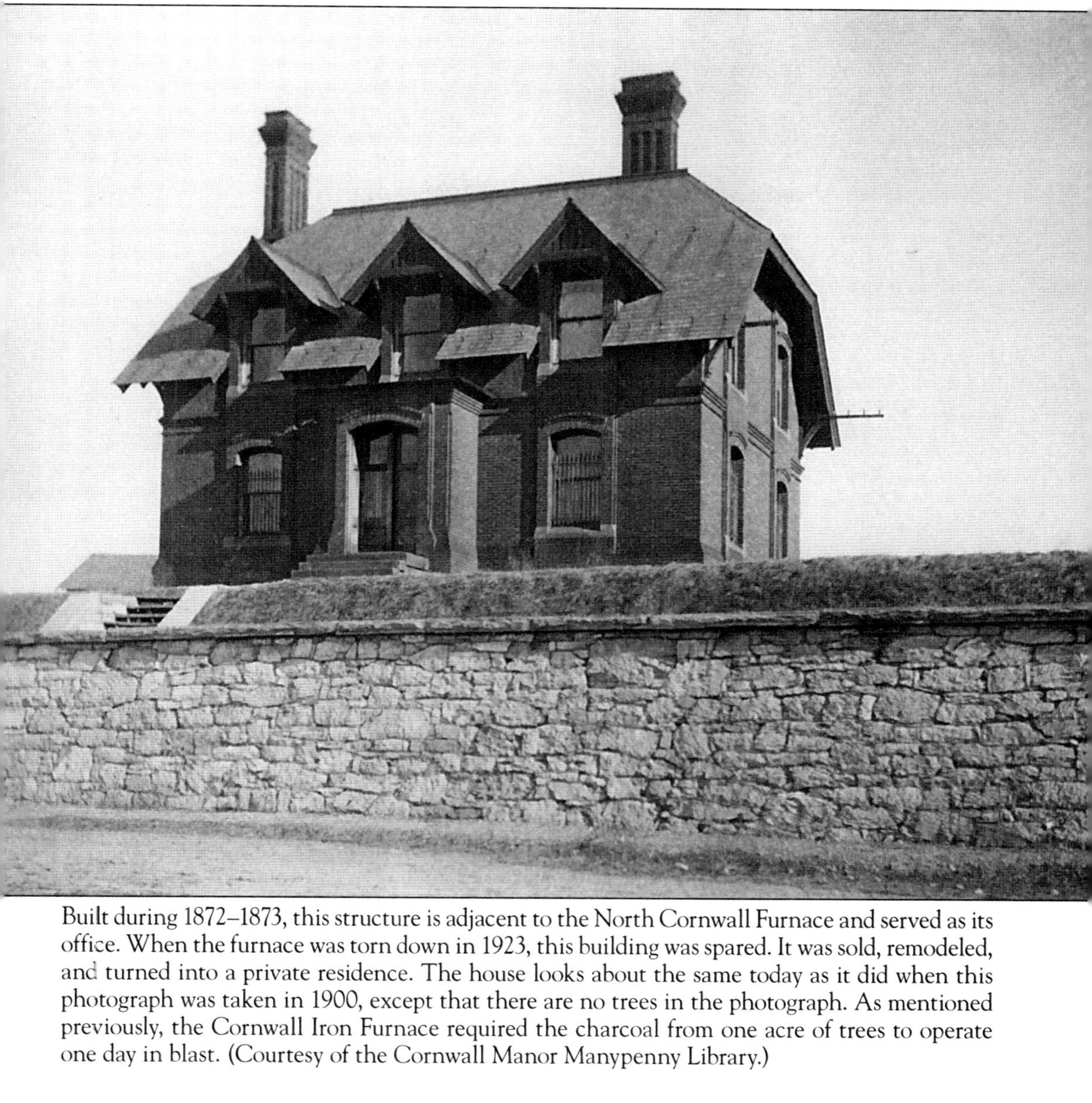

Built during 1872–1873, this structure is adjacent to the North Cornwall Furnace and served as its office. When the furnace was torn down in 1923, this building was spared. It was sold, remodeled, and turned into a private residence. The house looks about the same today as it did when this photograph was taken in 1900, except that there are no trees in the photograph. As mentioned previously, the Cornwall Iron Furnace required the charcoal from one acre of trees to operate one day in blast. (Courtesy of the Cornwall Manor Manypenny Library.)

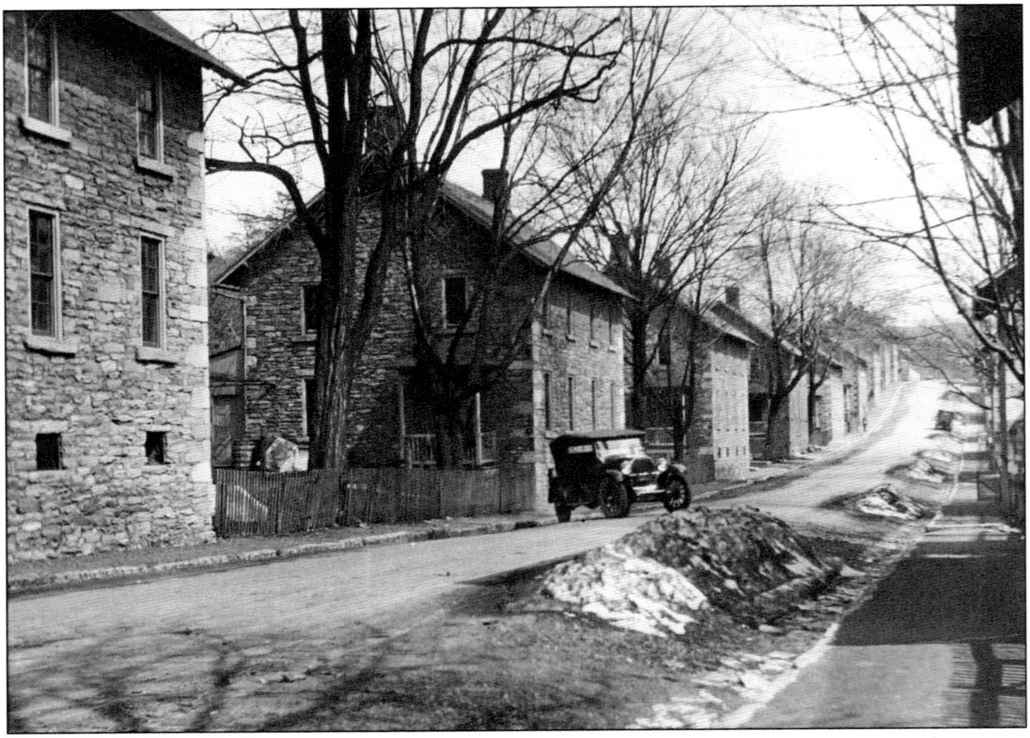

Minersvillage was established by the Cornwall Ore Bank Company between 1864 and 1900. It is the largest of the villages built for workers and is adjacent to the mine. Located on Boyd and Granite Streets, the houses are built of limestone, red sandstone, brick, or wood. The picture below shows the brick houses located on Granite Street. All of these homes were originally constructed for the miners. (Below, courtesy of Daniel Chernich.)

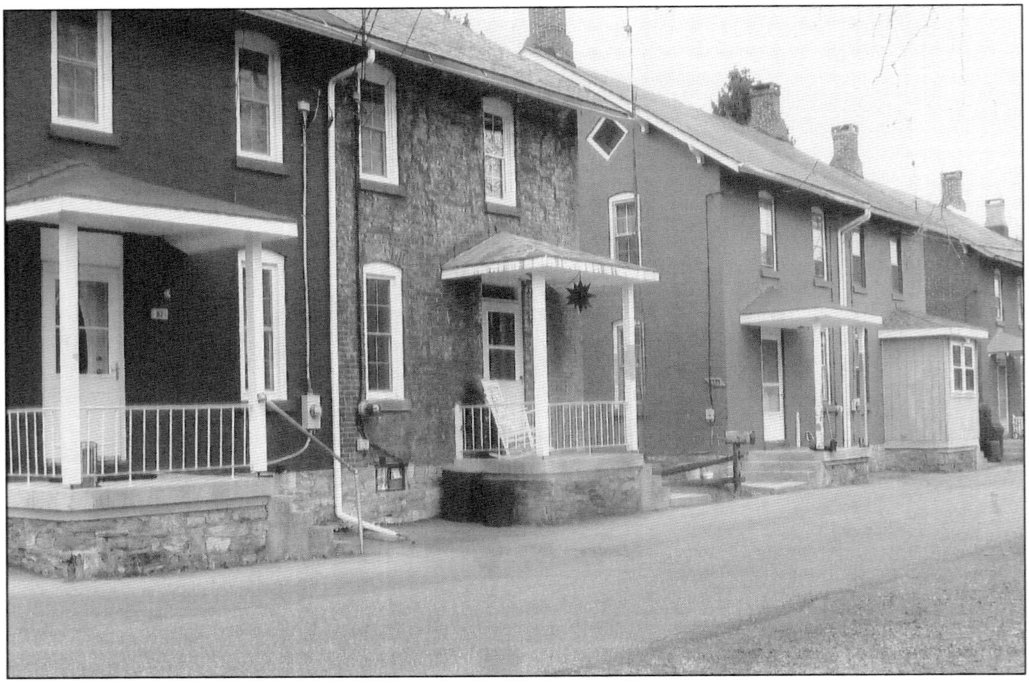

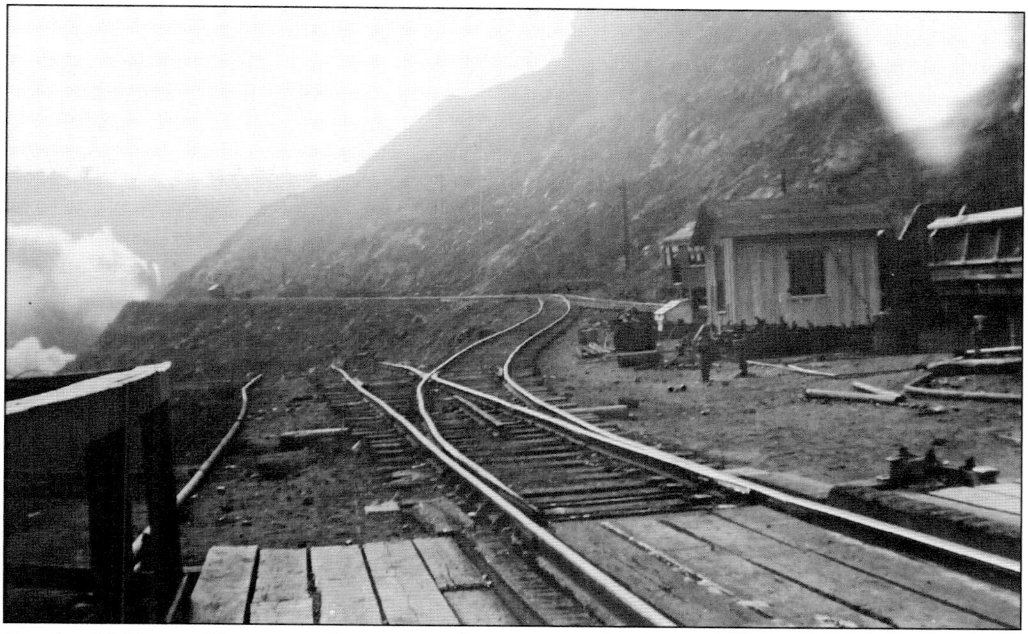

South of Minersvillage was the center of operations for the Cornwall Ore Bank Company Railroad. This image shows the tracks leading to the open pit. These rails were also used by the Cornwall Railroad to transport ore for processing and passengers who entrained at the Minersvillage Station.

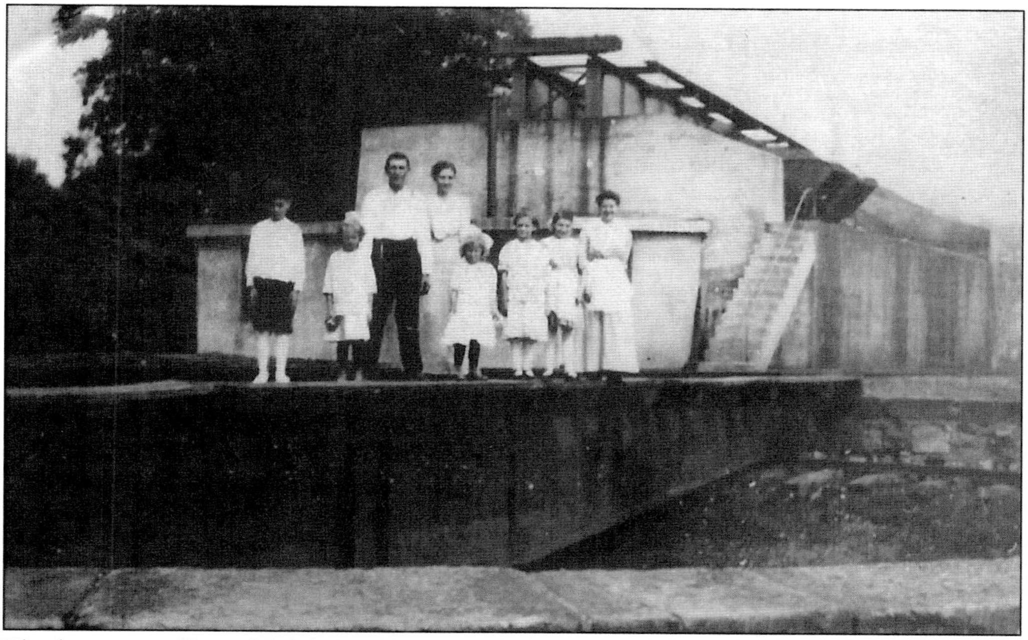

This large turntable enabled the engines to return to the engine house. At the rear of the turntable is a soft-coal bunker with an incline leading to its top. Railcars filled with coal were backed up the incline for unloading coal into the bunker, which had chutes located above the steps. When these were opened, coal would spill into the coal tenders connected to the engines. The engine house is not pictured but contained maintenance and blacksmith shops to repair the engines and keep the railroad line running. (Courtesy of Daniel Chernich.)

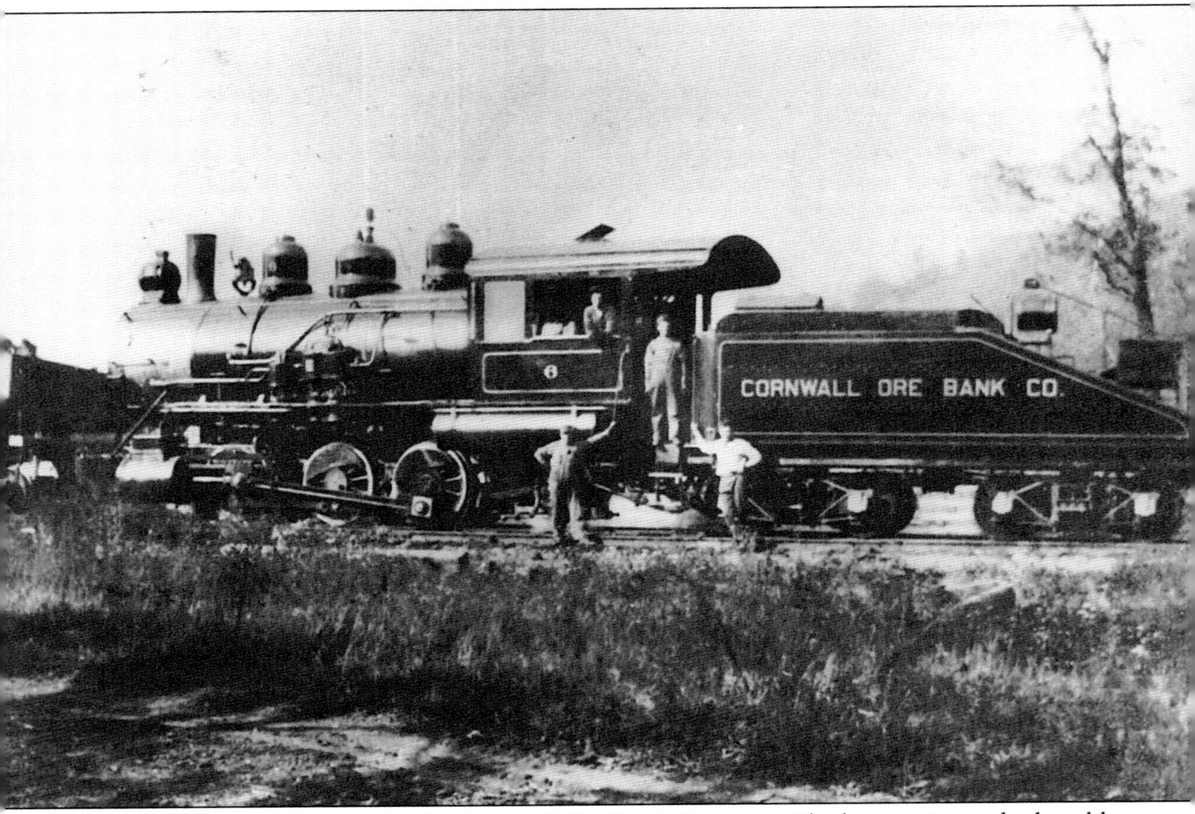

This engine was operated by the Cornwall Ore Bank Company. The locomotive and others like it operated only in the open-pit mine. Trains would haul ore out of the pit for processing before it was shipped to a furnace for smelting. They would also haul limestone and other waste rock to disposal areas. The Cornwall Ore Bank Company was formed in 1864 as a result of a lawsuit that began in 1851 and was finally settled in 1869. The suit was initiated because the various owners could not precisely account for the ore that was individually mined. At this time, 96 parts of ownership were established. J. Taylor Boyd was put in charge of all Cornwall Ore Bank Company mining operations except for the portion mined by the Robesonia Iron Company.

The buildings on this page were just west of Minersvillage. This photograph shows the Cornwall Ore Bank Company office, which was adjacent to the open-pit mine so the supervisors could view the operation from their windows. Notice the bell tower on the building. The bell would ring to signal the beginning and end of each workday and lunch period. After Bethlehem Steel purchased the mine, the administrative work was moved to a different office. This building was then turned into a residence, and Bethlehem Steel used a whistle to indicate the starts and ends of workdays and lunch periods. (Courtesy of Daniel Chernich.)

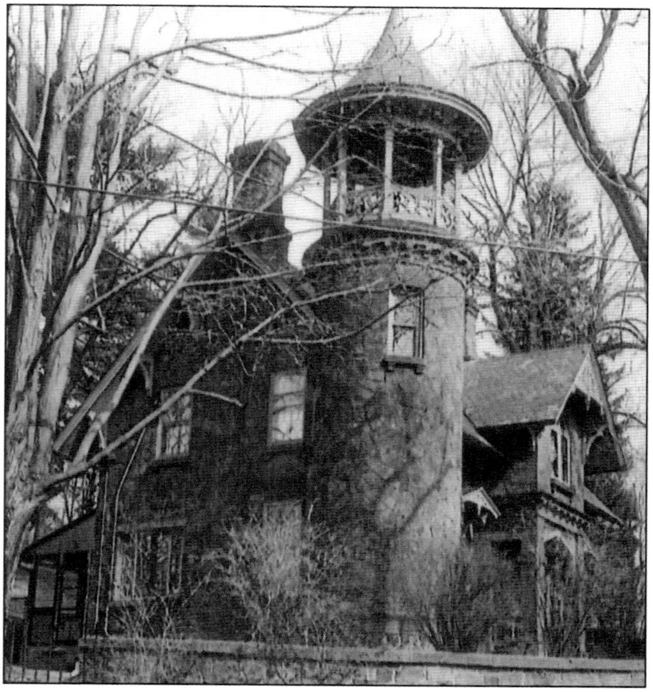

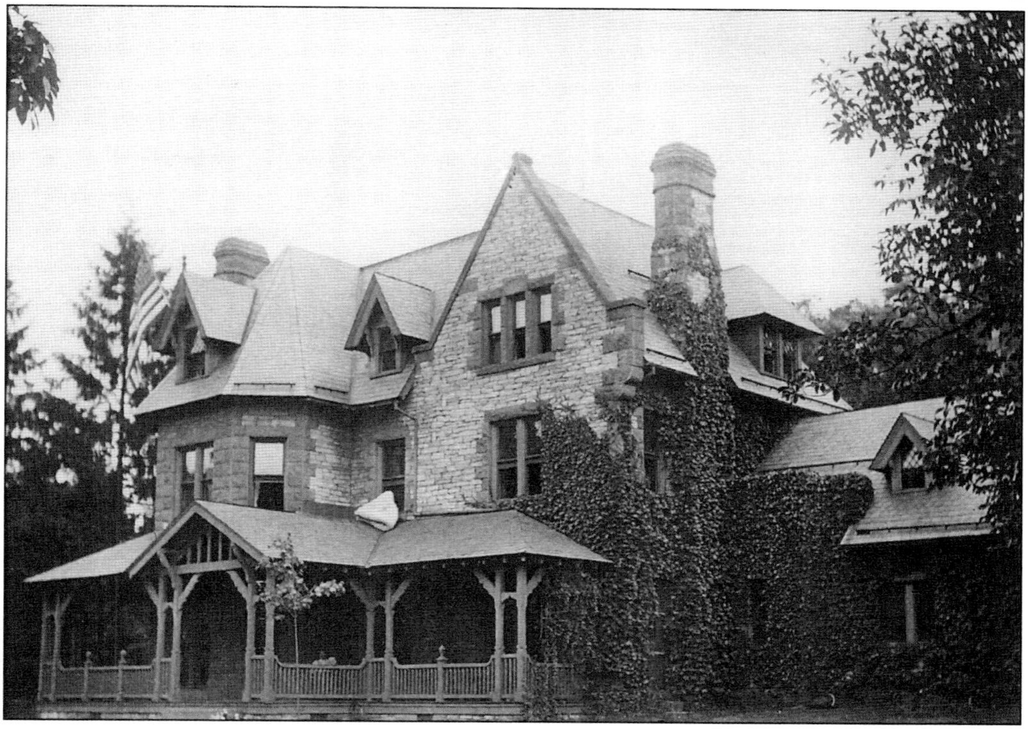

This is the residence of the mining superintendent for the Cornwall Ore Bank Company. The first mine superintendant to occupy the house after the Cornwall Ore Bank Company was formed in 1864 was J. Taylor Boyd. In 1964, the road leading to Minersvillage had to be relocated to expand the open pit to mine the east-end ore body. Both structures shown on this page were torn down at this time. (Courtesy of Cornwall Manor Manypenny Library.)

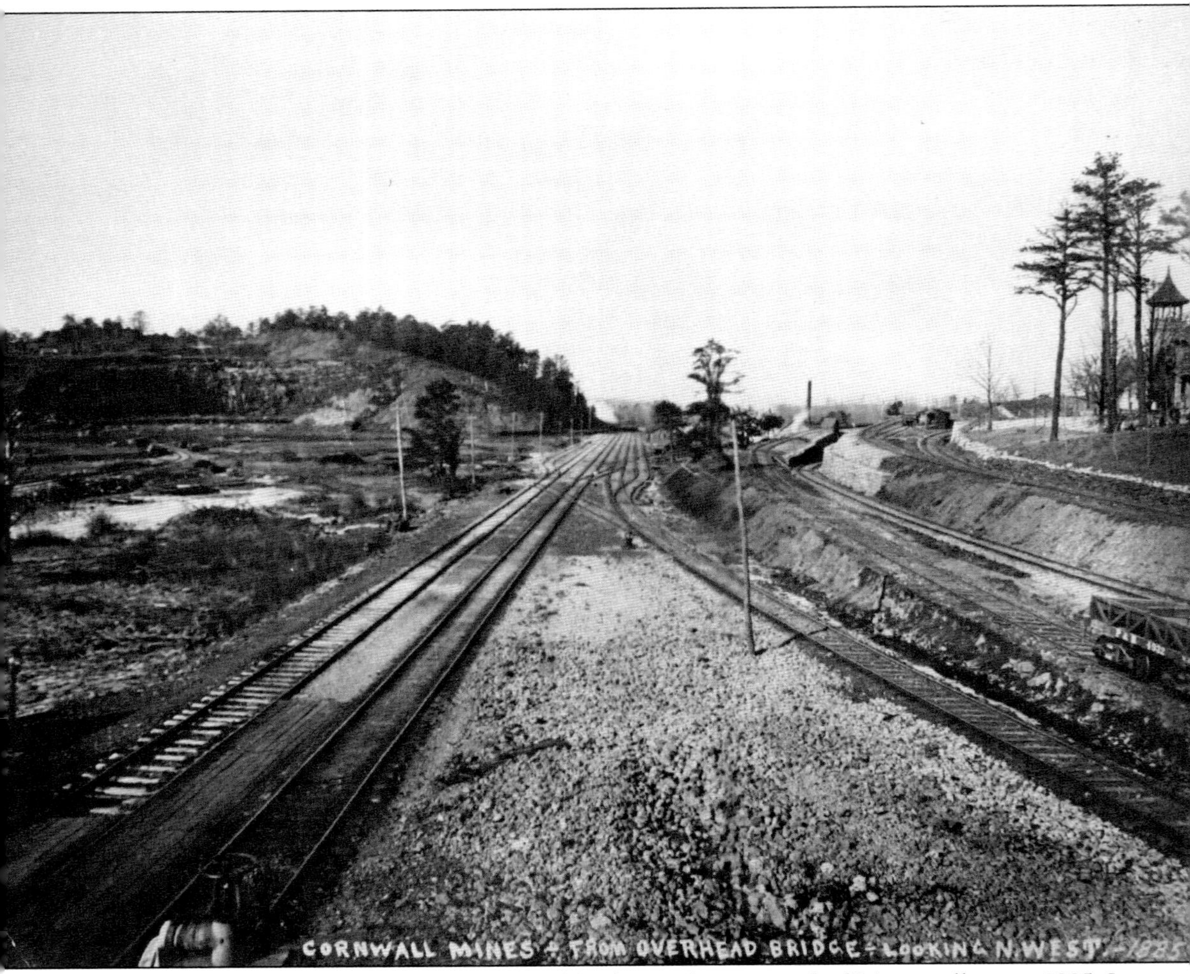

This picture was taken from the overhead bridge at the west end of Minersvillage in 1885. It shows the Cornwall Ore Bank Company office on the right, the road going past the office toward Minersvillage, and the train tracks from the Cornwall Railroad and the Cornwall Ore Bank Company Railroad. On the left side of the tracks is where Middle Hill and Grassy Hill would have been located. At this time, the company had started to remove the hills to get down to the ore. This is the beginning of the large open-pit mine that produced ore until the 1950s.

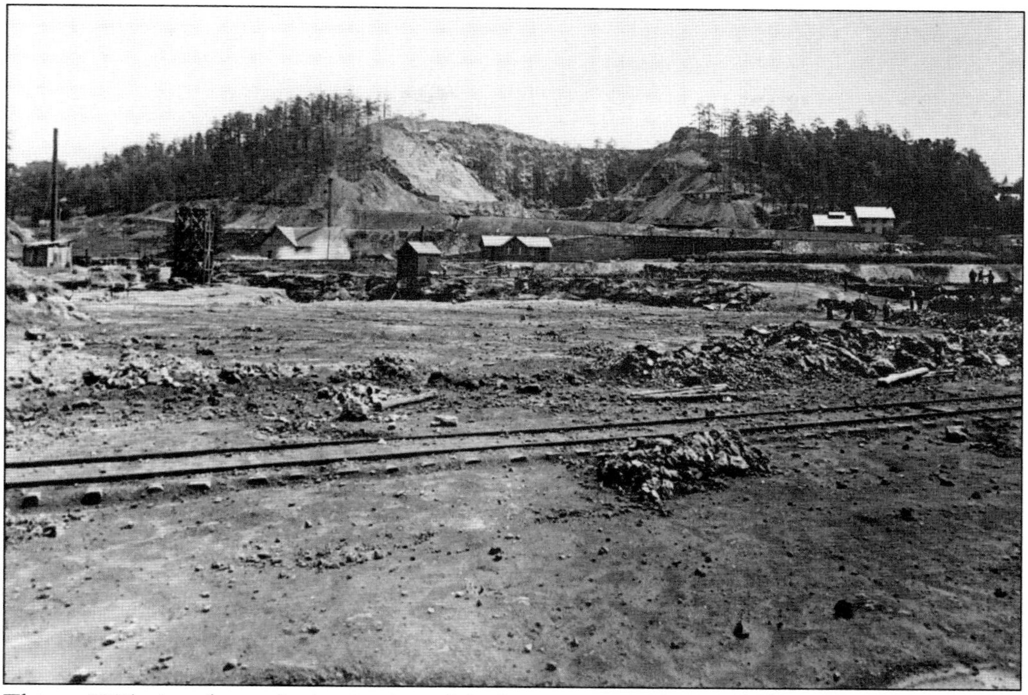

This c. 1885 view shows the beginning of the large open pit from a different direction. On the right edge of the image is the top of the tower of the Cornwall Ore Bank Company office. In the background is Big Hill, the area where the first mining took place. The gap in the hill is where the mine was located. The ore deposit in Big Hill was nearly exhausted, and the mining operation would soon be moved to another ore deposit. At the right of the photograph, a mule team and cart and a group of men can be seen working. The railroad tracks used to haul away waste material are shown in the foreground.

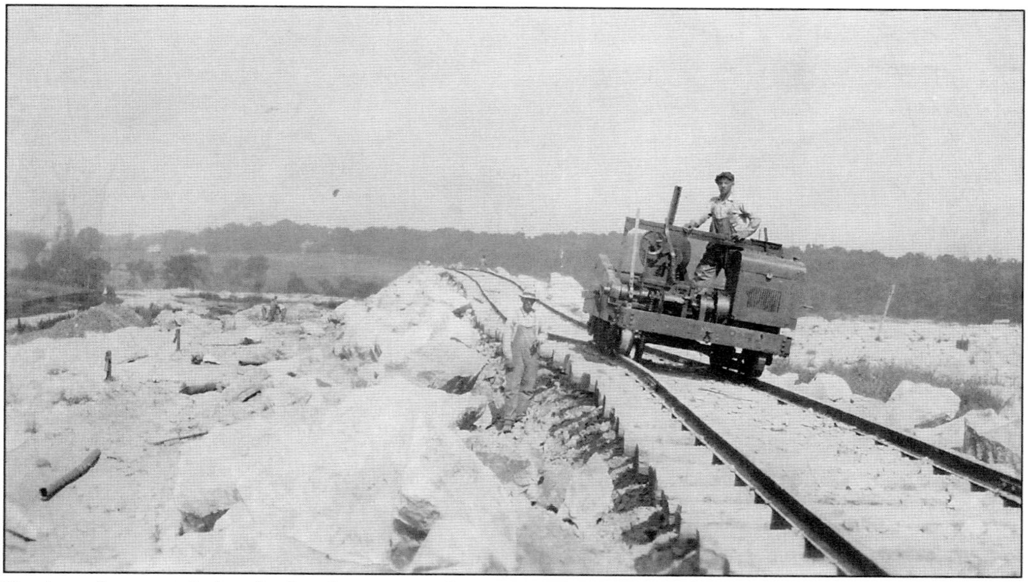

Rock and material that had to be removed to get to the ore was shipped by railcar to a dumpsite. The picture shows one of the waste disposal areas and the rail shifter, which was able to shift the rails laterally. The railcars were side-dumping.

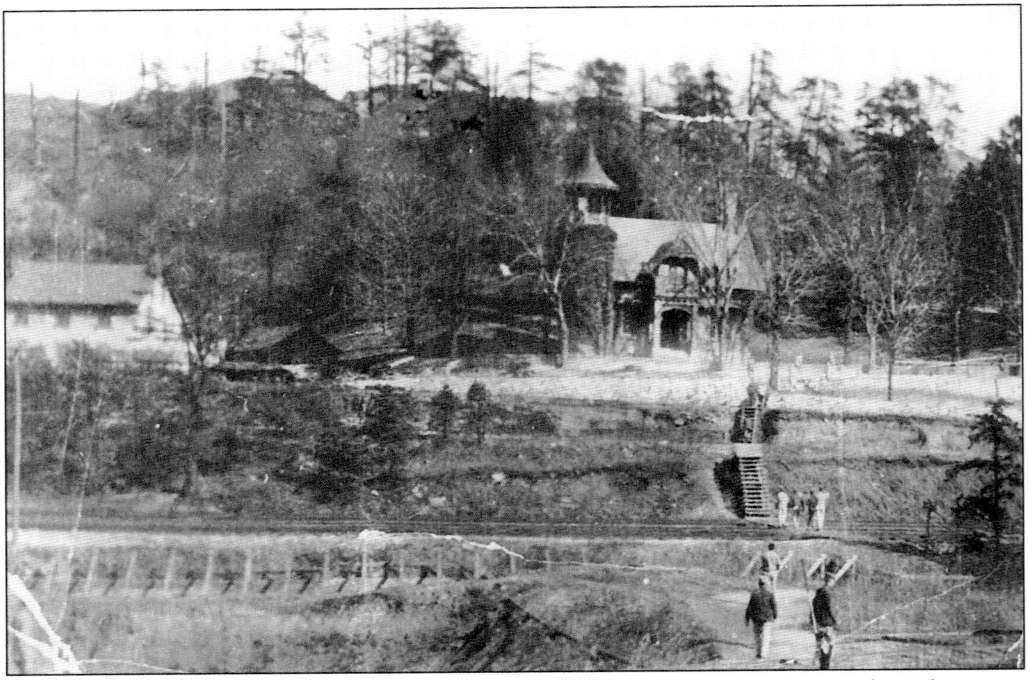

This view shows an early stage of the development of the large open-pit mine. At the right center of the picture is the Cornwall Ore Bank Company office, with its bell tower. The bell rang at the beginning of each workday and again at the end. The stairs allowed the workers to enter and exit the mine from the office where the time clock was located. (Courtesy of Daniel Chernich.)

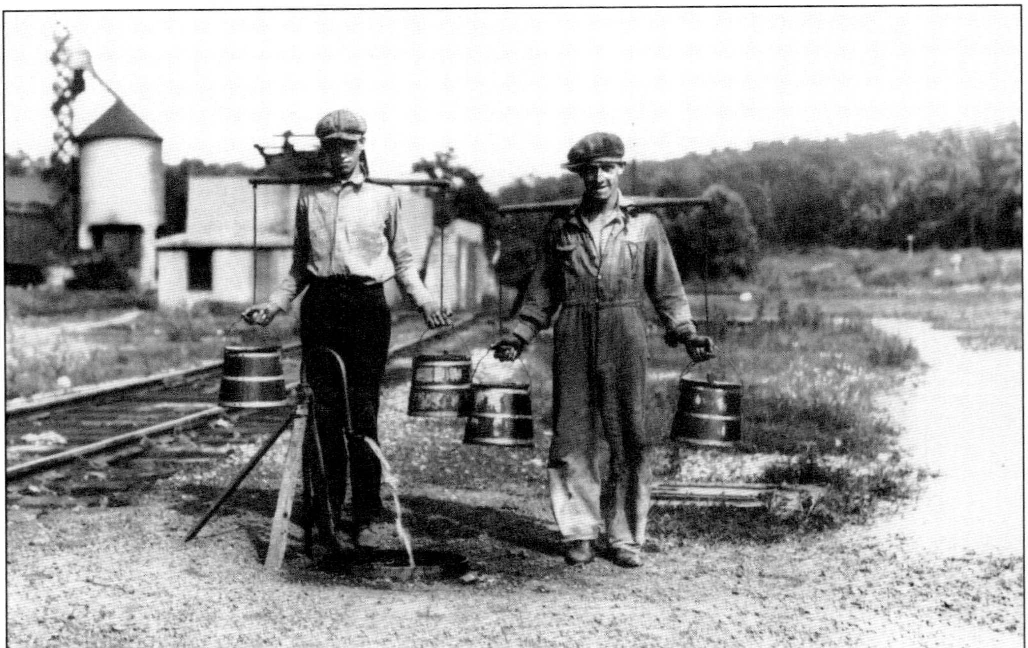

These unidentified laborers worked as water boys. All day, they toiled in the mines carrying drinking water to the miners. The census from 1880 lists children as young as 12 years old working as day laborers in the mine.

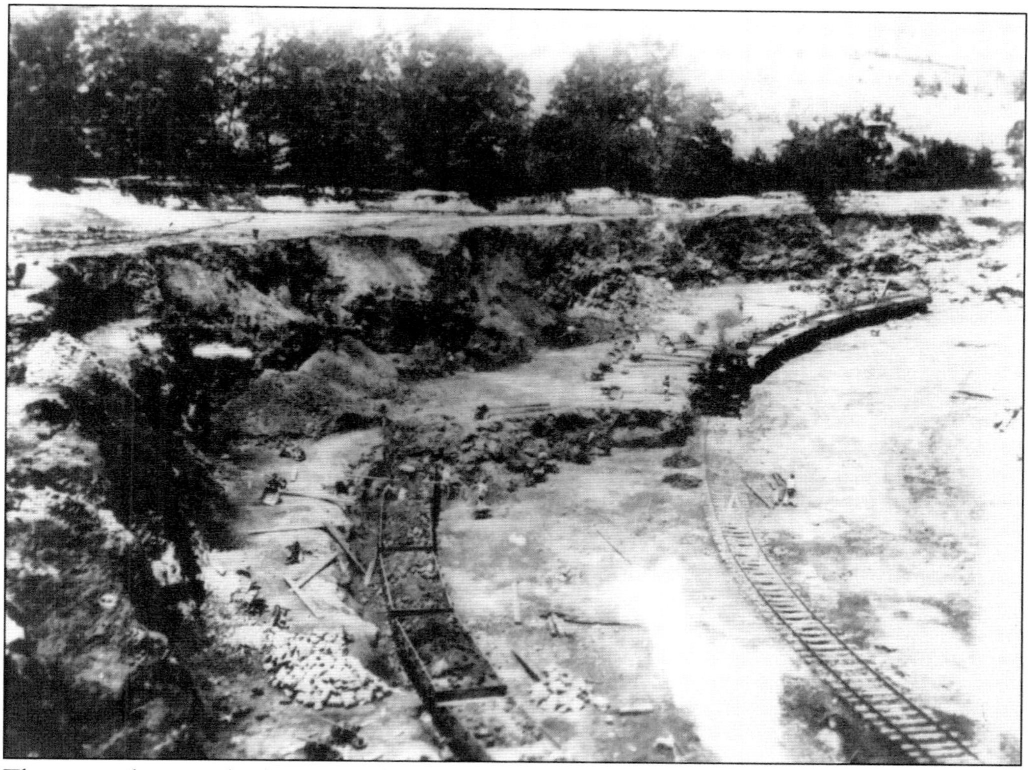

These two photographs show an early method of mining known as the wheel-and-bench method. It consisted of working the ore or other material in 20-foot benches and hauling it in wheelbarrows to railcars. The image above shows the elevation of the benches. They were the same height as the tops of the cars. Planks were laid across the gap between the bench and train to allow wheelbarrows access to dump their loads into the railcar. When the cars were full, the engine would pull the train away, and the cars would be emptied as shown in the photograph below. This mining method was used until steam shovels were purchased to load railcars in 1910.

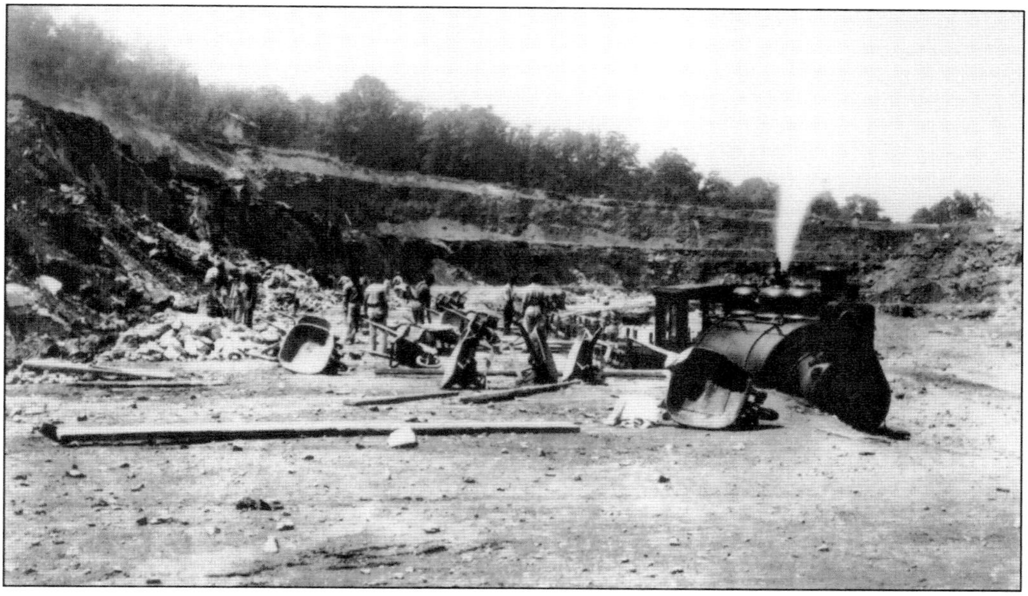

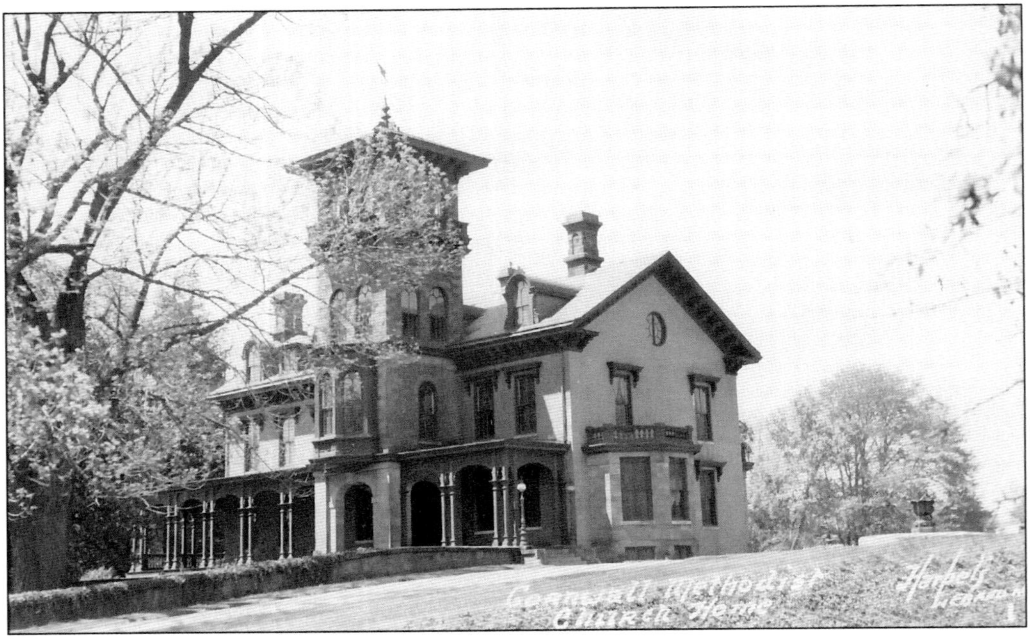

This mansion was built around 1773 by Peter Grubb's son Curttis. The original structure had 19 rooms and overlooked the original charcoal furnace and the meadow that was used to test shoot the cannons supplied to the Continental army during the Revolutionary War. Robert Coleman gained ownership of the mansion in 1798. In 1865, Coleman remodeled the home, adding 10 more rooms and a formal garden. The photograph above shows the front of the house. The picture below depicts part of the formal garden with walking paths and a pond. The mansion now contains seven apartments for residents of Cornwall Manor. (Both, courtesy of Cornwall Manor Manypenny Library.)

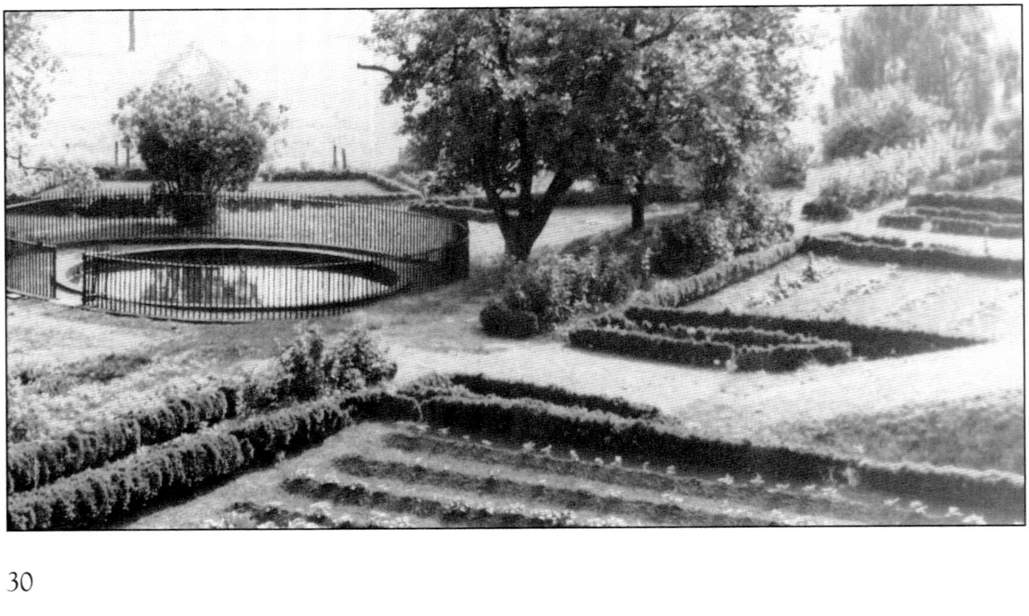

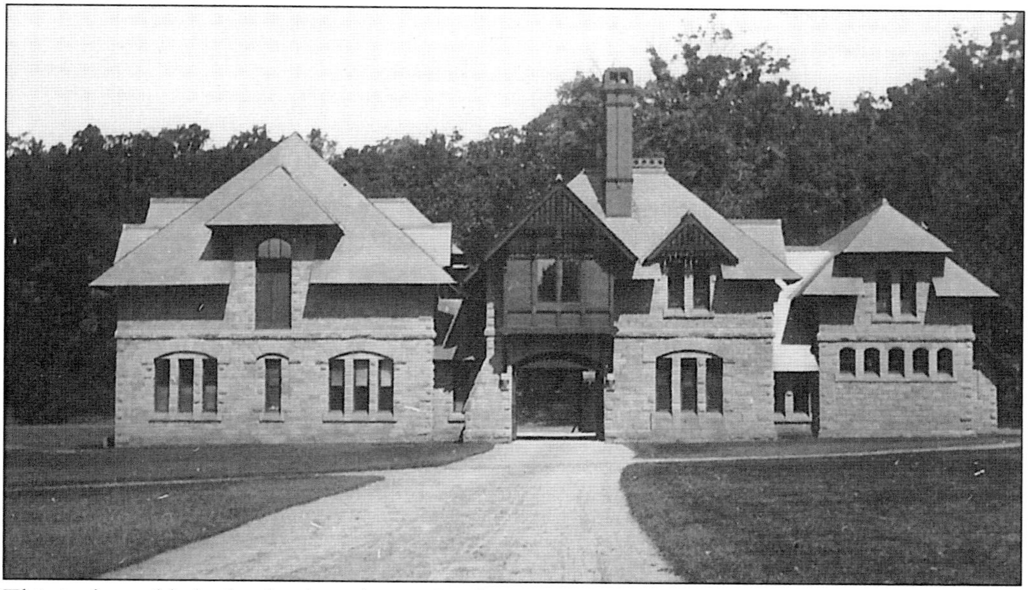

This is the stable built of red sandstone for the iron master's horses. It was located very close to the iron master's mansion pictured on the preceding page. Cornwall Manor turned the stable into apartments. (Courtesy of Cornwall Manor Manypenny Library.)

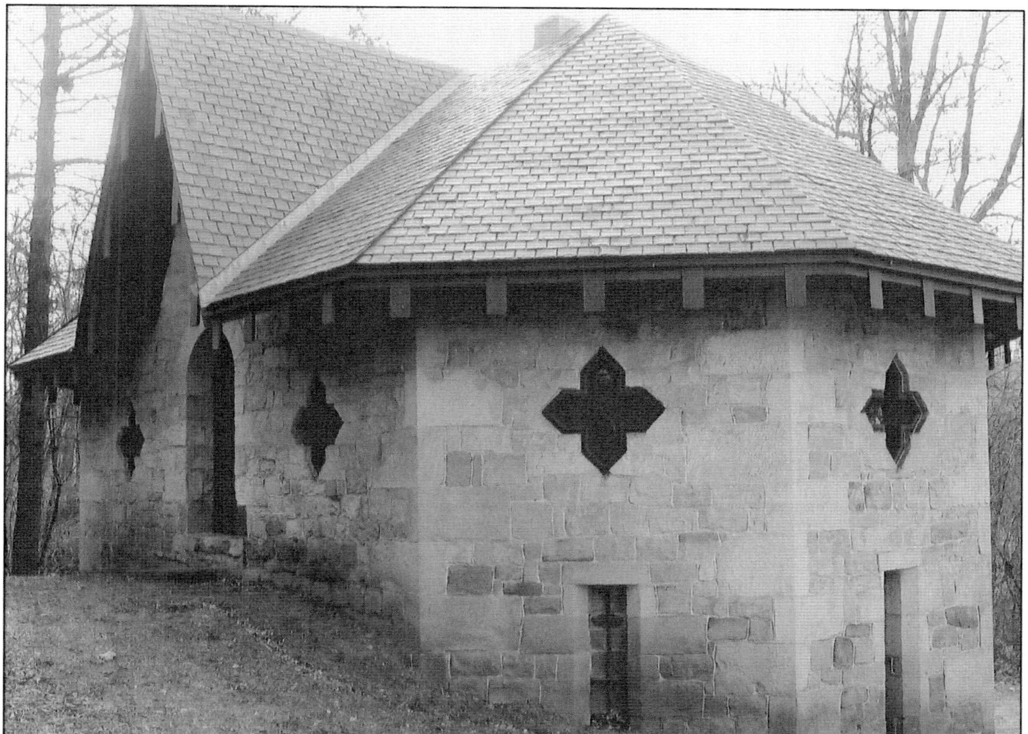

This picture shows the abattoir, where local farmers would bring their animals for slaughter. The meat would then be sold for food. With its farms, mills, stores, and butchers, the iron plantation was nearly self-sufficient. This stone building is now part of the Cornwall Iron Furnace State Historic Site.

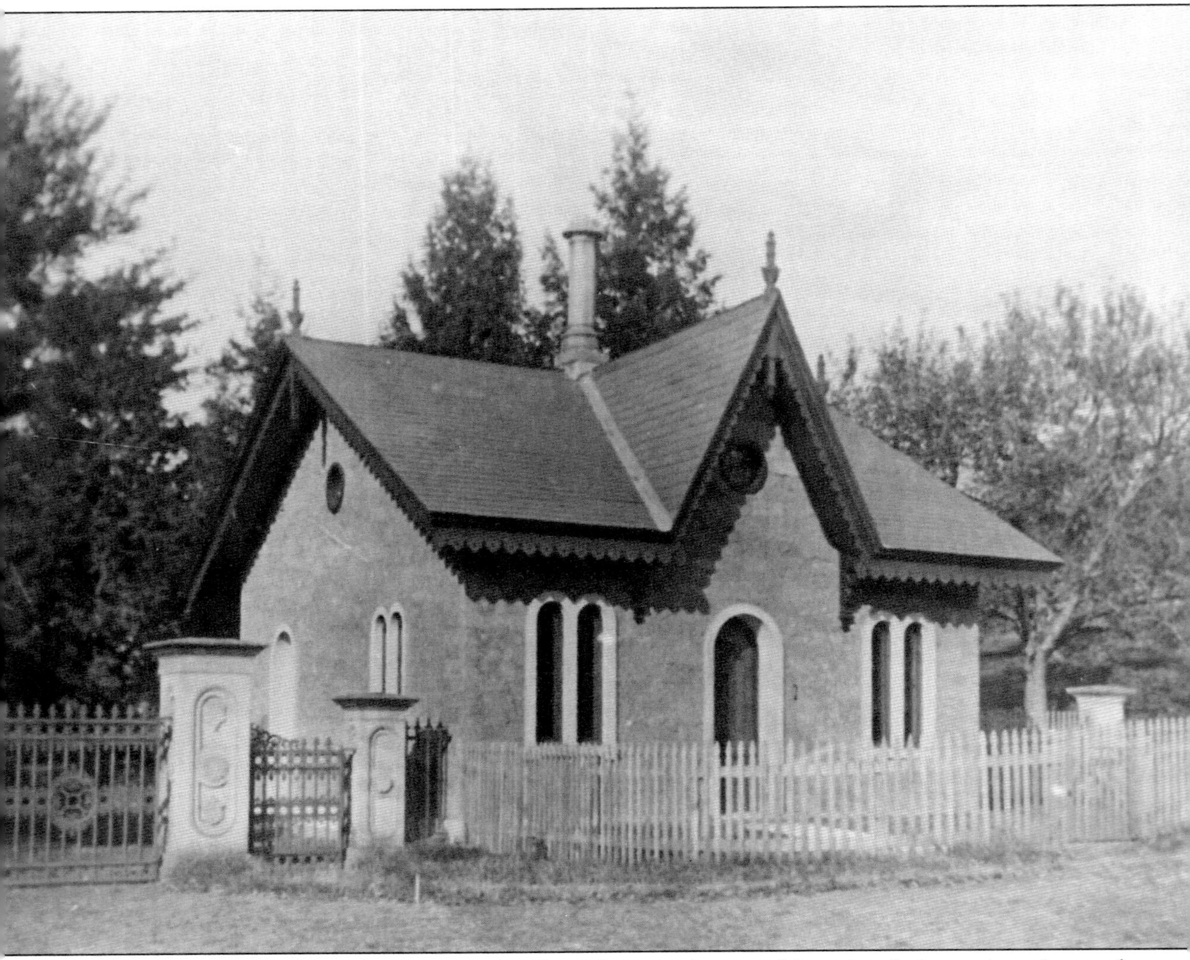

This photograph shows the paymaster's office, where workers would receive their earnings. Located across the street from the charcoal furnace, the office was built along the short private road leading from the mansion to the furnace. A very large safe is still inside the building. (Courtesy of Cornwall Manor Manypenny Library.)

Three

Steam Shovels Come to Cornwall

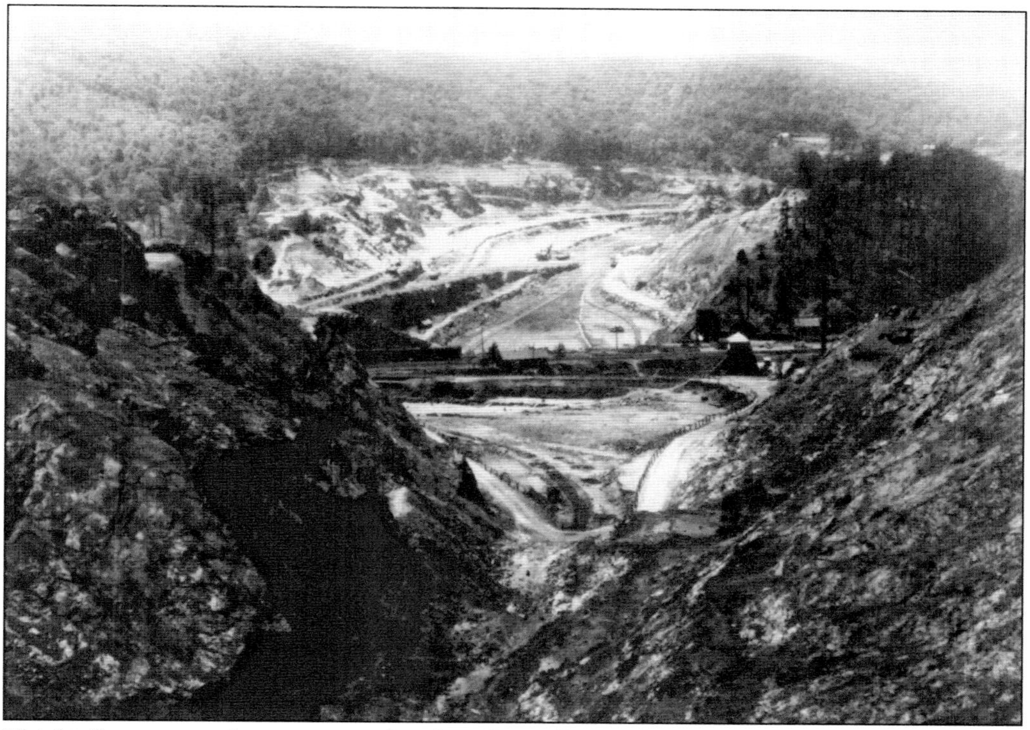

This bird's-eye view of mining was taken from Big Hill around 1906, where the first open-pit mine was started. The Big Hill ore body was exhausted, and the larger open-pit mine was underway near Middle Hill and Grassy Hill. As seen in this photograph, the larger open pit is beginning to get some depth to it. A steam shovel and train tracks can be seen in the distance.

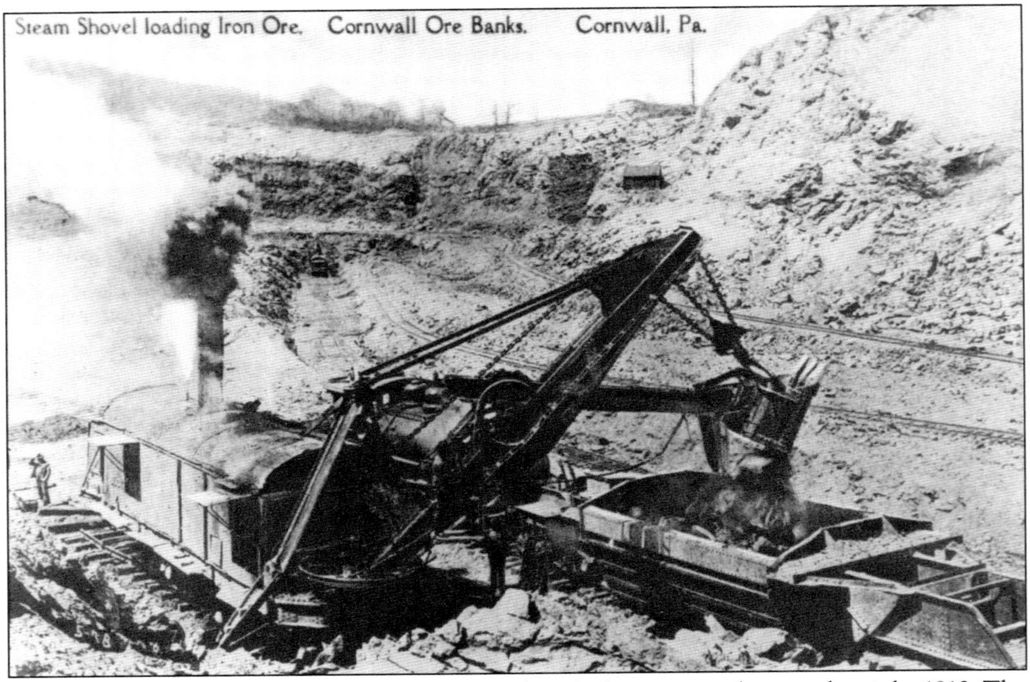

These photographs show one of the three steam shovels that were working in the pit by 1912. The shovels had a dipper capacity of 2.5 cubic yards, meaning each could hold about seven tons of ore. They operated on standard-gauge railroad track, each with a crew of 11 men. This crew included one steam shovel operator, one craneman, five laborers, two men who operated steam-powered jackhammers, one blaster, and one man who was tasked with carrying water and keeping the shovel supplied with coal. The five laborers kept the shovel pit level and placed ties and rail for the next move. The steam jackhammer was used to create holes in the rock. The blaster would fill the drilled holes with dynamite and blast the rock into pieces that were small enough for the steam shovel to handle. A close look at the image below reveals a boom, which was used to support the rotating shovel. It is unusual in that only the boom swivels, rather than the main body.

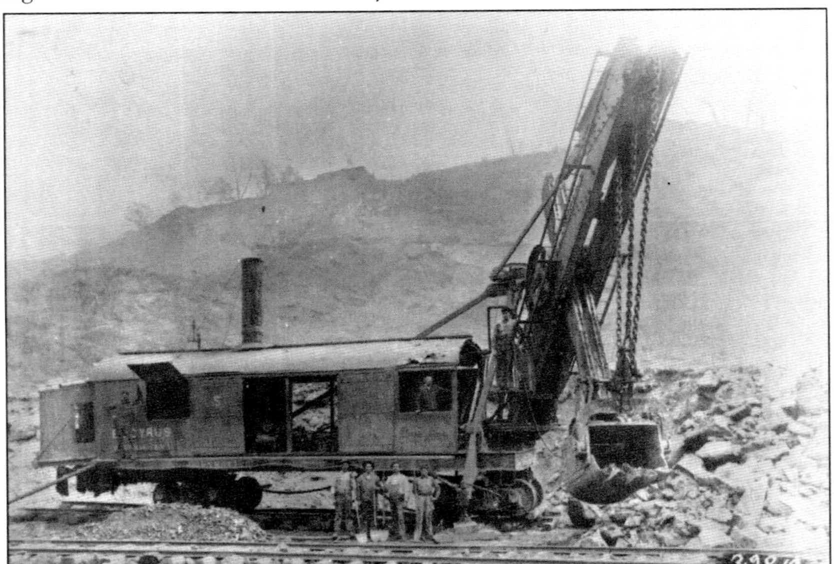

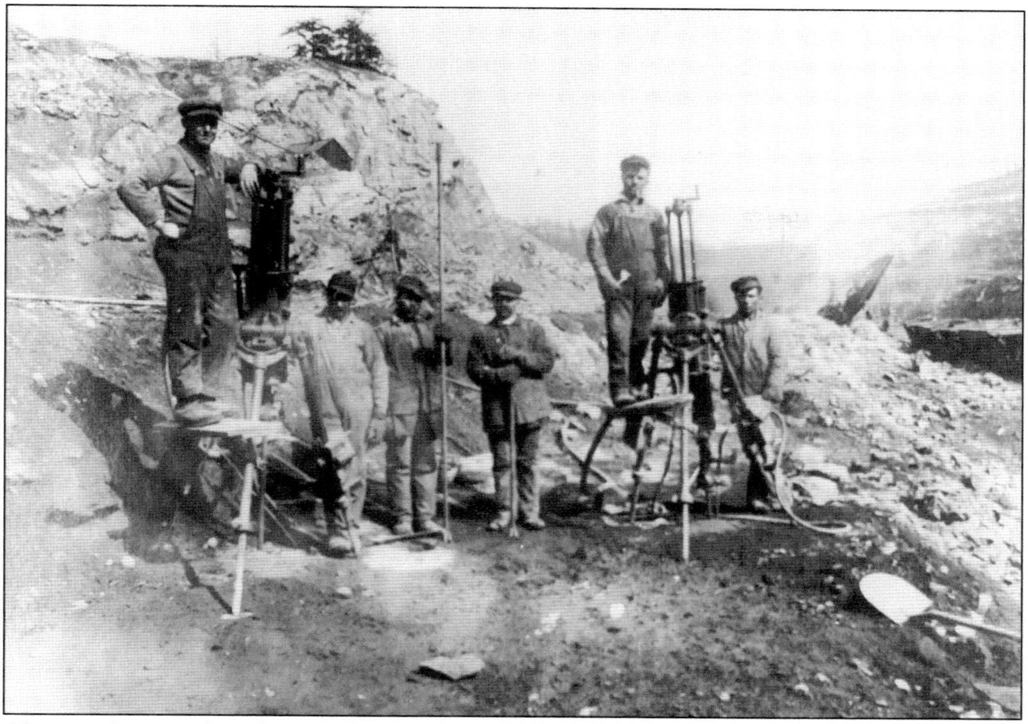

Along with steam shovels, men operating steam-powered drills worked ahead of the steam-shovel crew. The drill operators would make holes that were then filled with dynamite and blasted ahead of where the shovel was working. Most of the blasted chunks of rock would be small enough for the shovel to load into railcars.

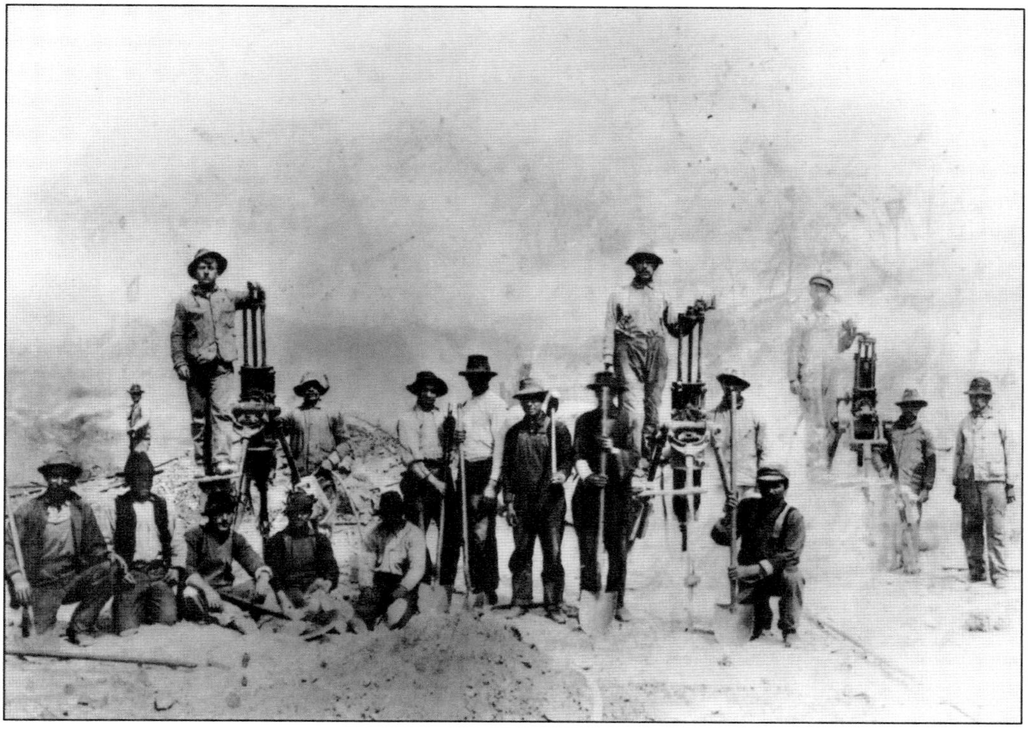

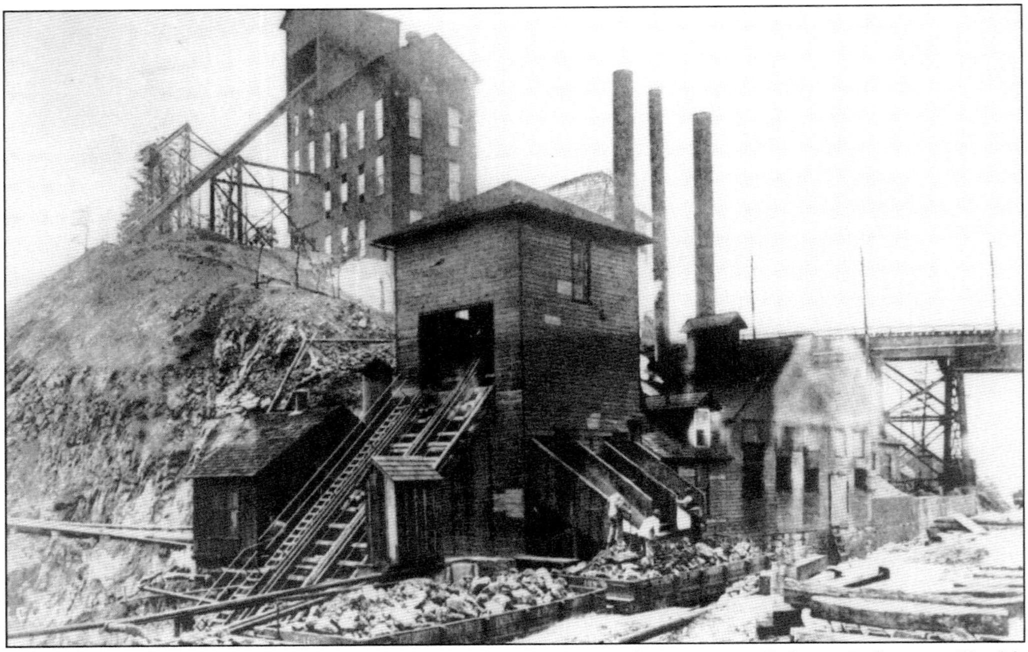

In 1786, Peter Grubb Jr. granted his one-sixth interest in the mine to Robert Coleman. Grubb reserved the right for himself and his heirs to obtain a supply of ore to supply one furnace. This one furnace would become the Robesonia Iron Company. The Robesonia Iron Company headframe is shown here. Workers are loading ore from the chutes into railcars. The Cornwall Ore Bank Company headframe can be seen in the background. The Robesonia Iron Company was acquired by Bethlehem Steel in 1926.

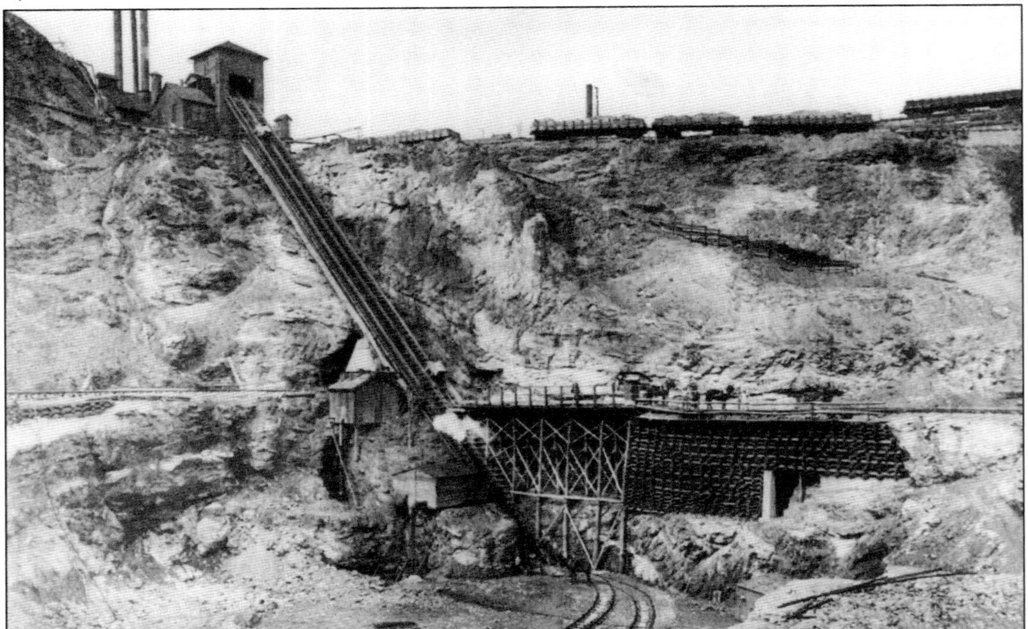

This picture from about 1910 shows the incline leading to the Robesonia Iron Company headframe. Skips can be seen on the incline. Railcars are at the top of the mine ready for loading.

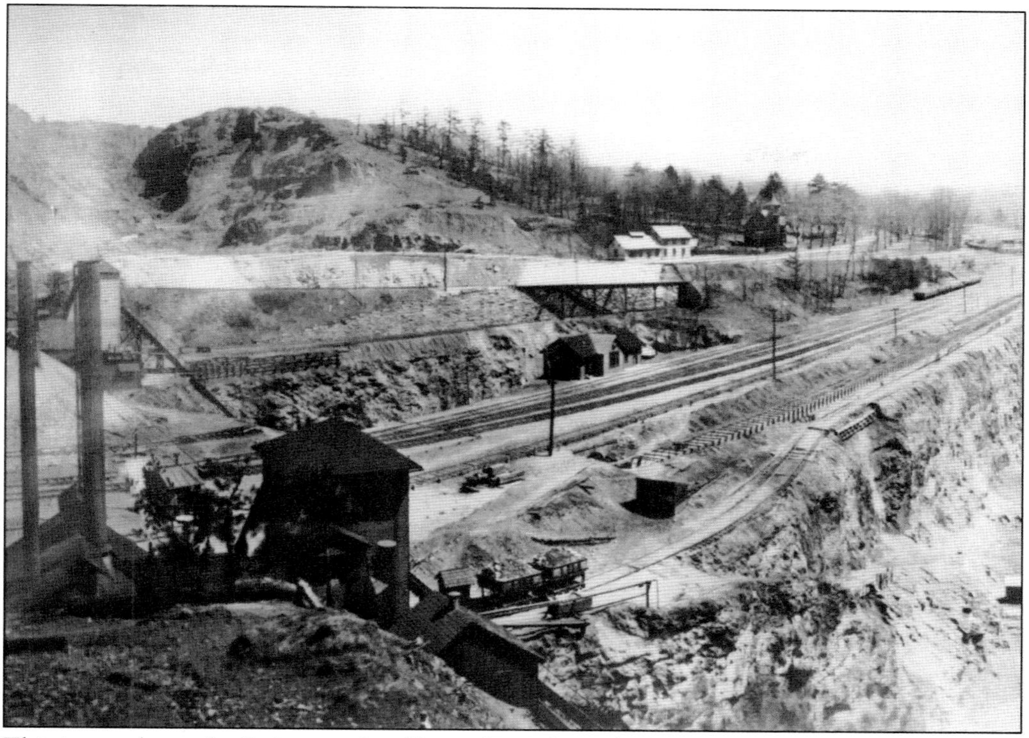

This image shows the Robesonia headframe from above, looking down on the railcars. Big Hill is also visible in this photograph, which also includes the road that goes past the Cornwall Ore Bank Company office on its way to Minersvillage. Extensive railroad tracks can be seen just below the road and near the headframe. Another mining operation is also visible. Both mines were part of the second, larger open pit. The pit was divided by the strip of land with the railroad tracks. This division was later removed so that the two mines became one.

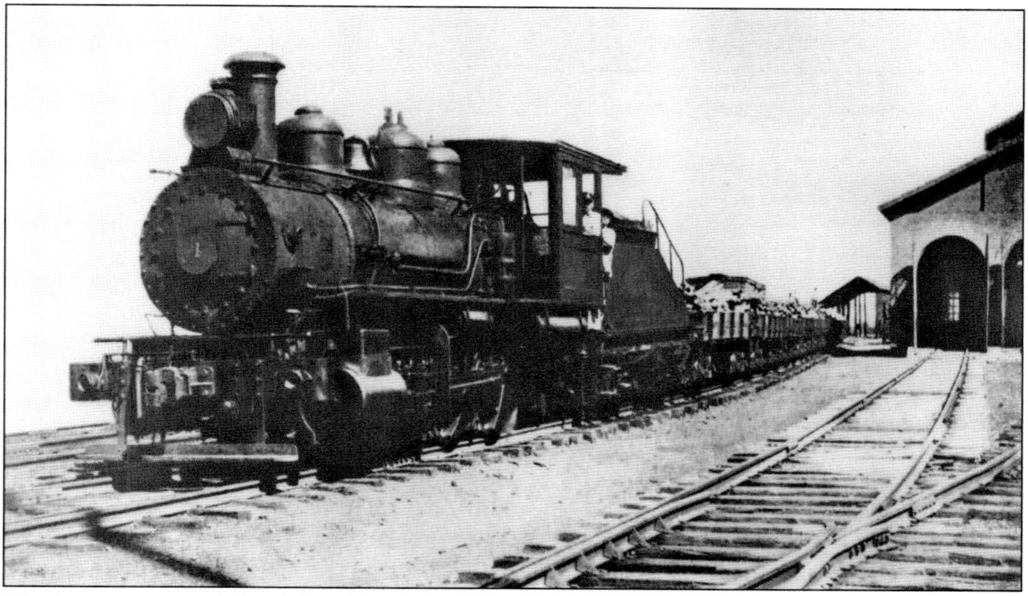

This locomotive with loaded railcars belonged to the Robesonia Iron Company. The borough of Robesonia is about 20 miles east of Cornwall in Berks County.

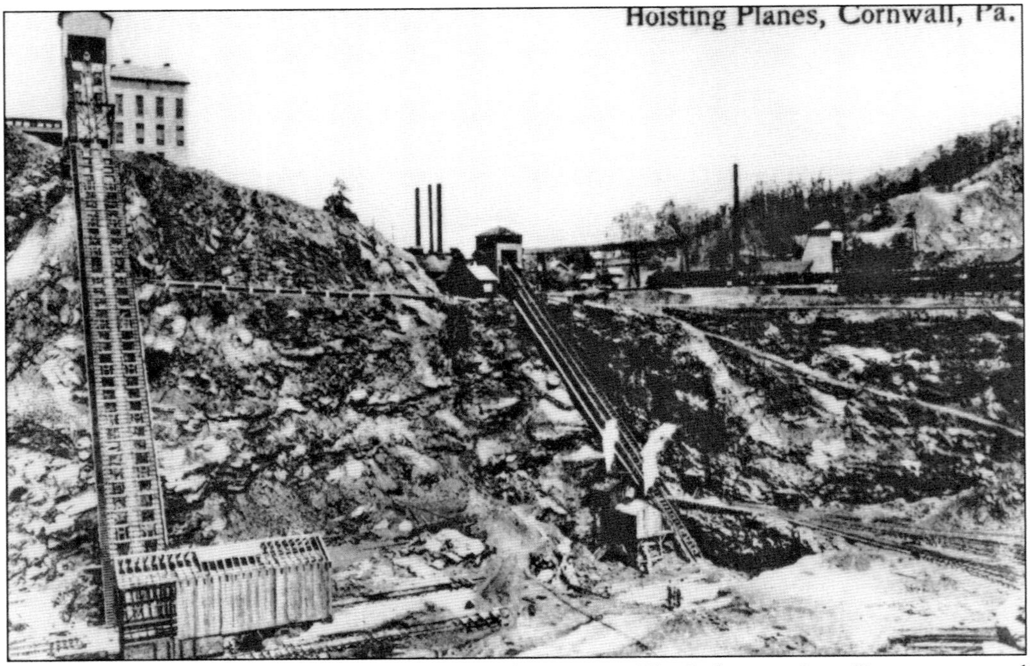

This photograph shows the hoisting inclines for the open pit. The Robesonia Iron Company owned the one on the right, and the Cornwall Ore Bank Company operated the one on the left.

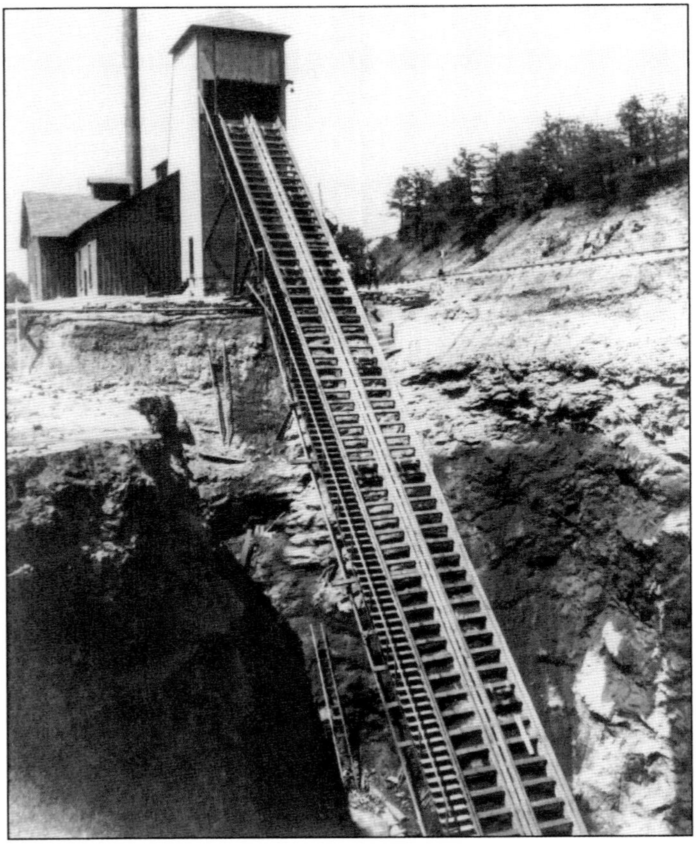

This is a picture of the Robesonia Iron Company's No. 1 Hoist. Material would be transported to the top of the hoist by steel cars known as skips. A cable pulled the skips to the top of the incline, where they were dumped. The ore and rock were crushed and separated.

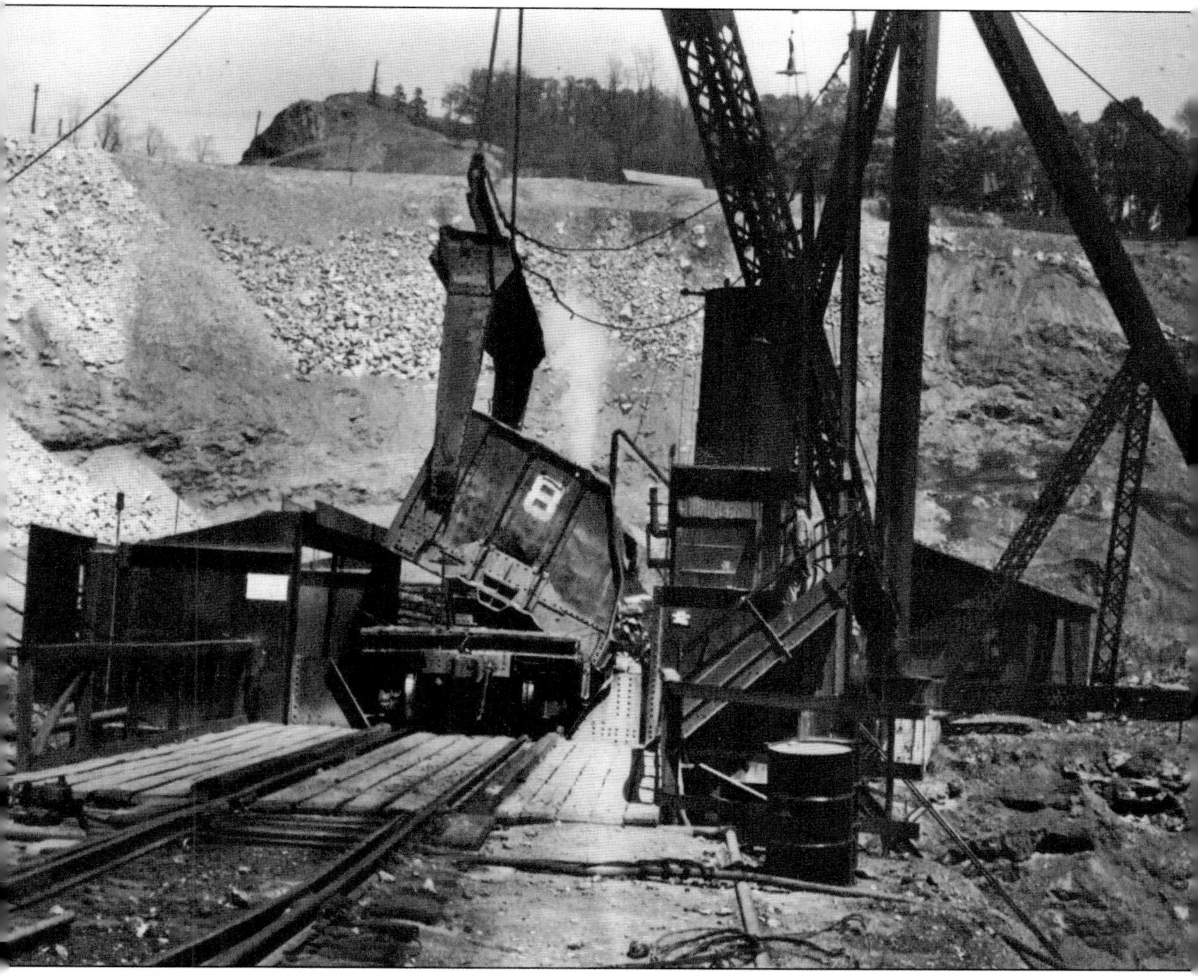

Steam shovels would load the railcars with raw material. These railcars would be pulled by the Cornwall Ore Bank Company steam engines to the bottom of the incline, where a dumping station was located. The side-dumping railcars would be emptied by the use of a crane that would lift the body of the railcar to dump its contents into a crusher. The crushed material was brought out of the pit in skips.

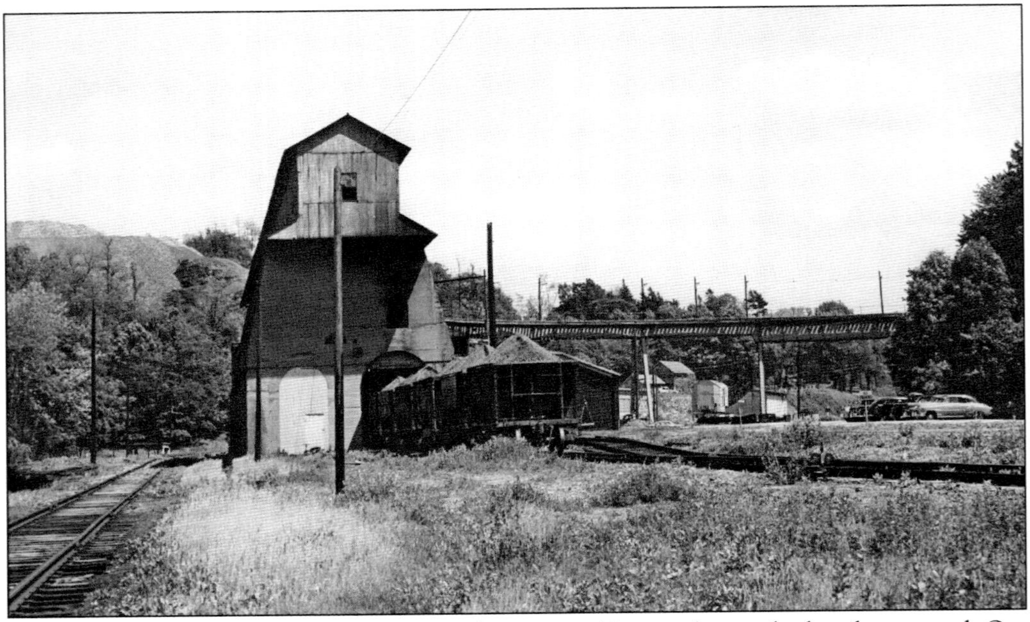

When the material was brought to the surface, it would again be crushed and separated. Ore would be taken to the loading pocket (above) for placing into railcars. It was hauled there by a transfer car (below) operating on an elevated track. Waste rock would be carried to one of the rock dumps by the Cornwall Ore Bank Company Railroad. The cars filled with ore would be pulled to the Lebanon Concentrator Plant by way of the Cornwall Railroad.

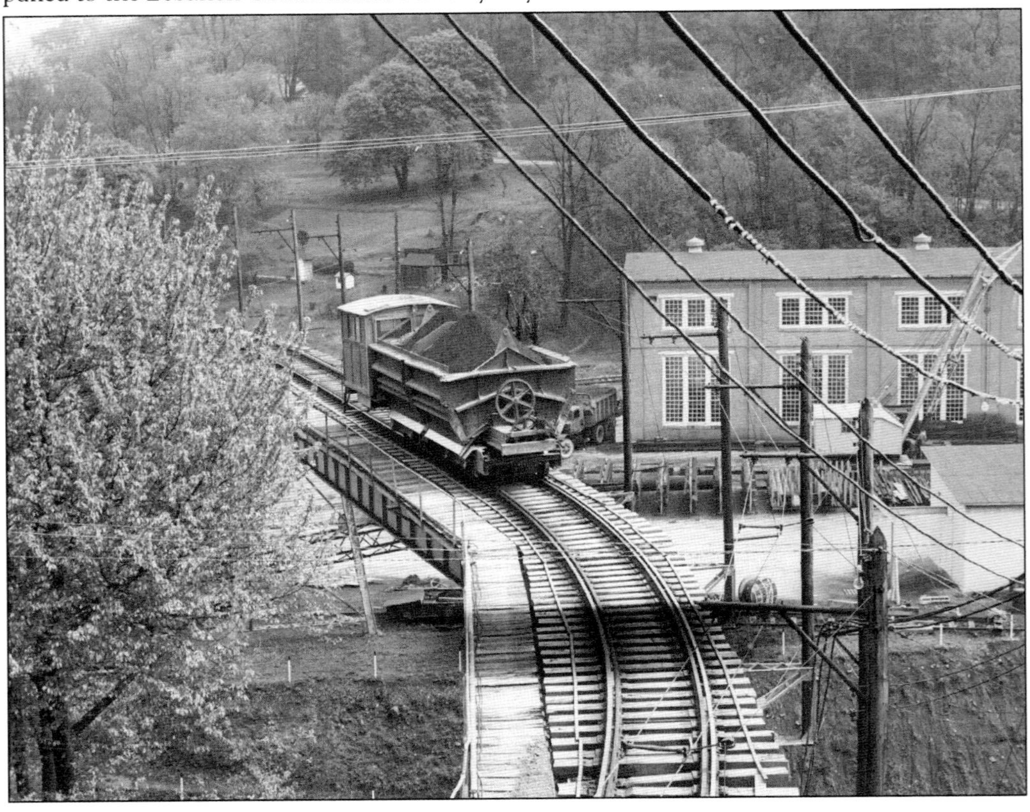

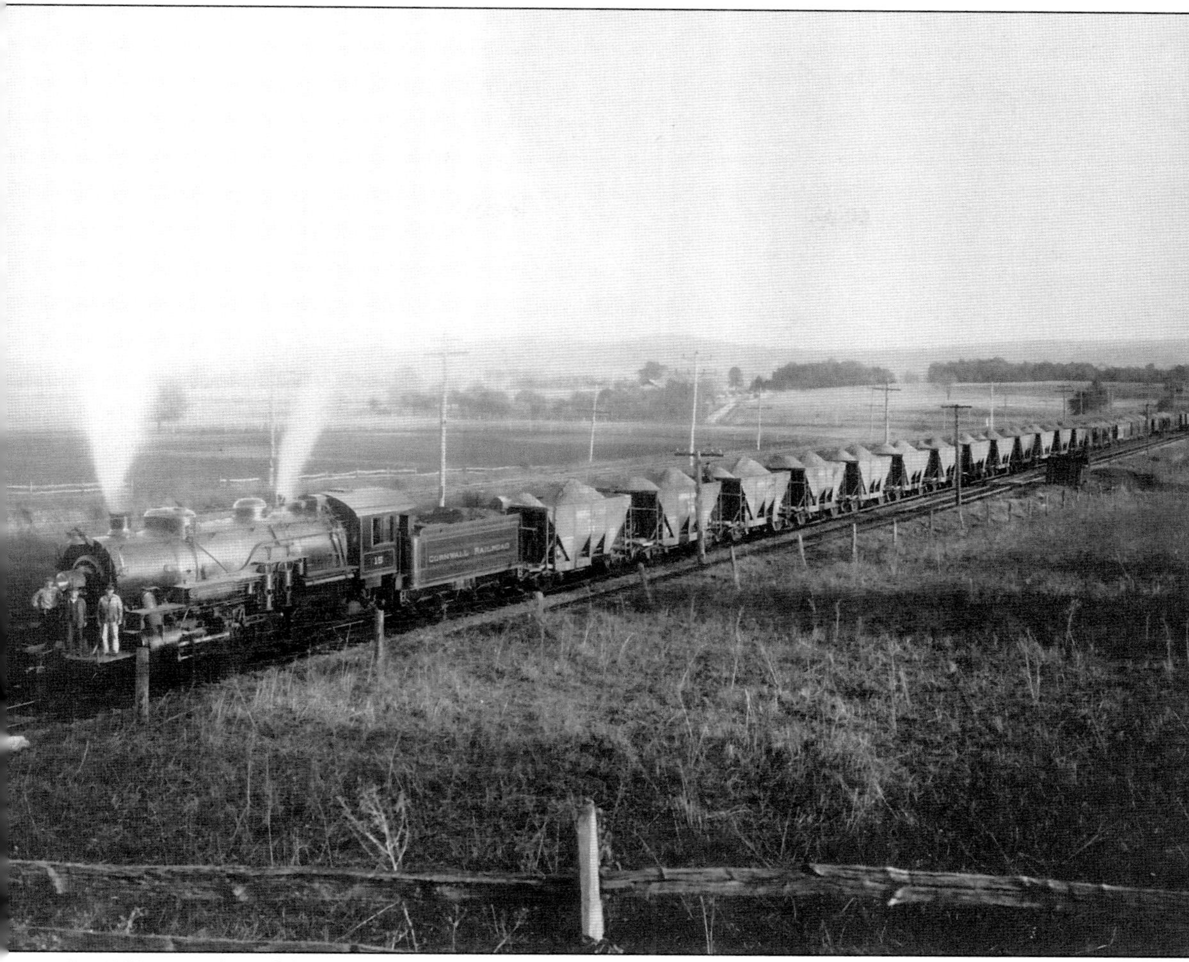

In 1850, a number of the members of the Coleman family incorporated the North Lebanon Railroad Company to haul ore from the mines at Cornwall to the Union Canal landings in Lebanon. All the stock of this company was taken by members of the Coleman family. The name of the railroad was changed to the Cornwall Railroad in 1870. In 1962, the railroad made its last trip hauling ore to the Lebanon Concentrator Plant when the pellet plant was completed in Cornwall. The railroad still hauled railcars containing pellets to Lebanon until October 1964, when the track was leased to and operated by the Reading Railroad.

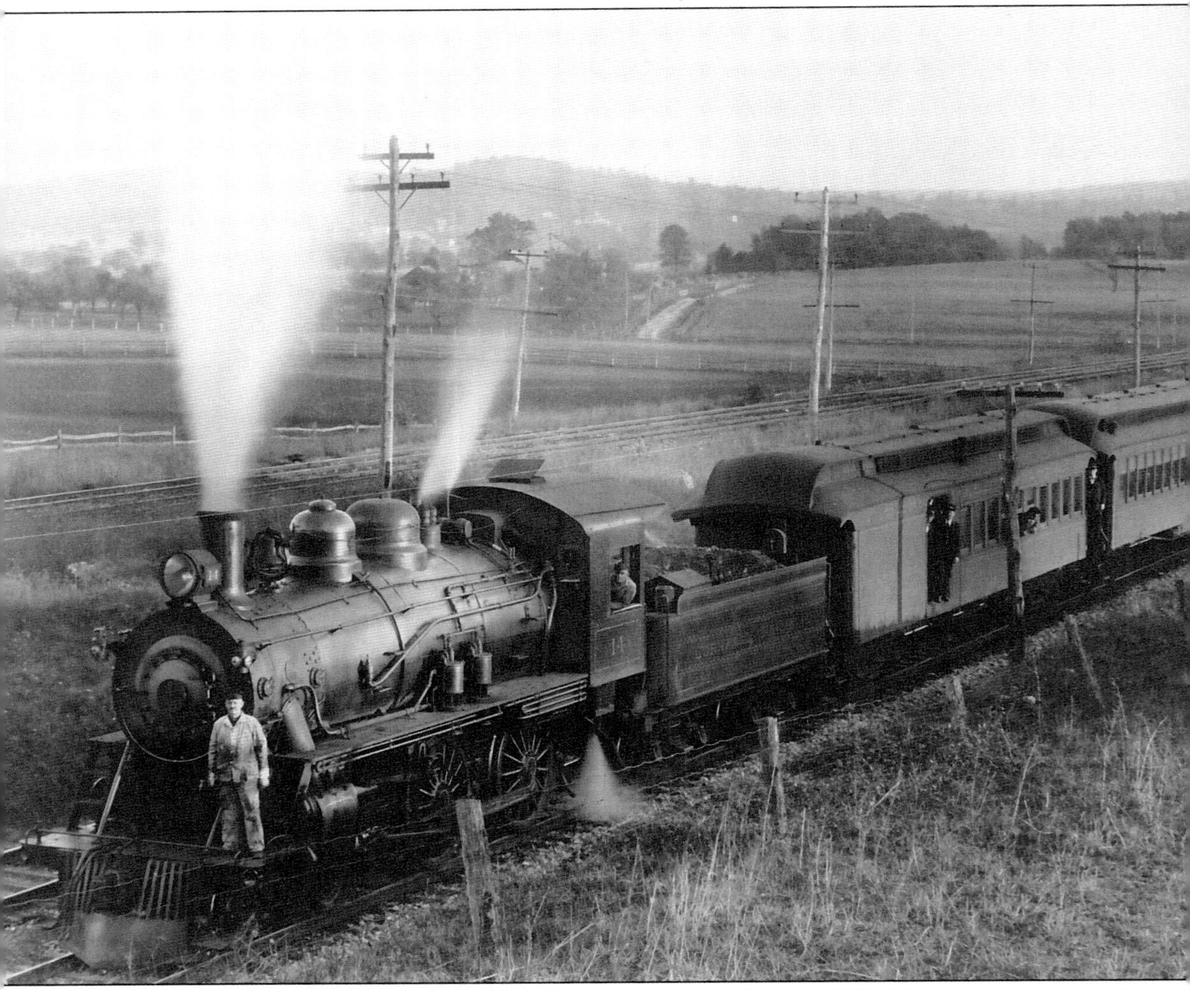

The Cornwall Railroad also operated passenger service until January 1931. The service connected with the Philadelphia and Reading Railroads in Lebanon, thereby enabling travel to and from Harrisburg, Reading, Pottsville, Philadelphia, and New York. The passenger service also connected to smaller local cities and towns.

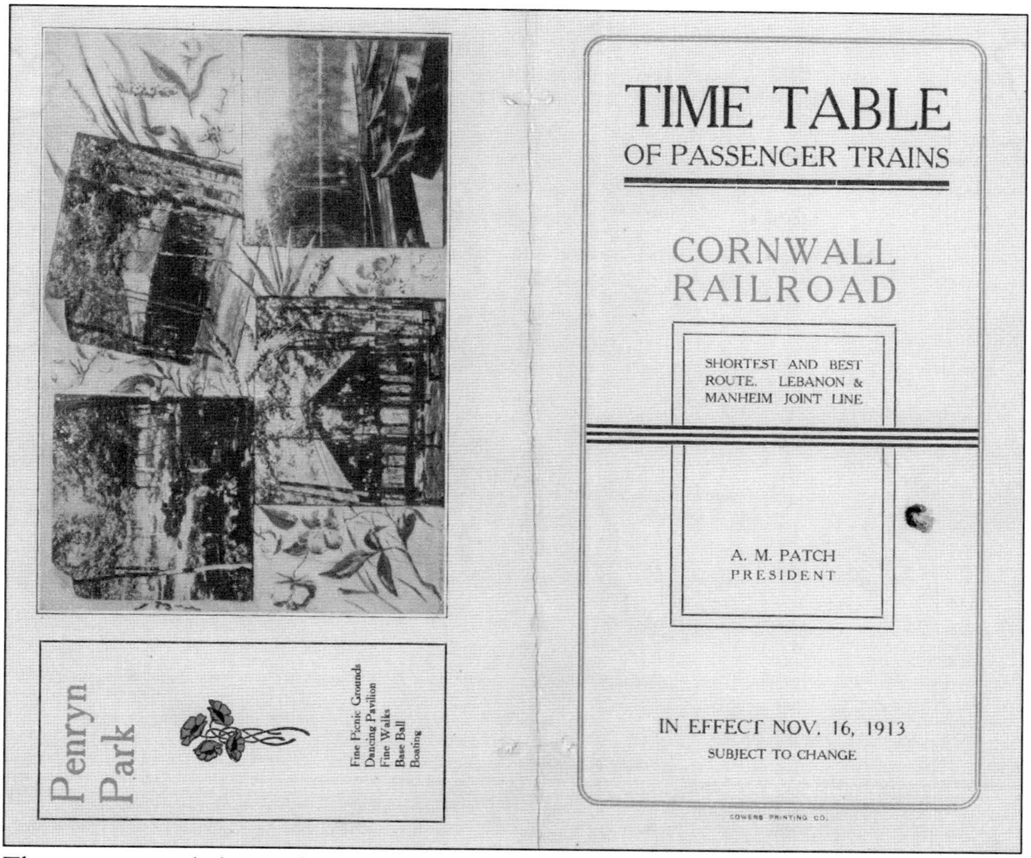

These images include an advertisement for Penryn Park (above) and a train schedule (below), with the times at stations served.

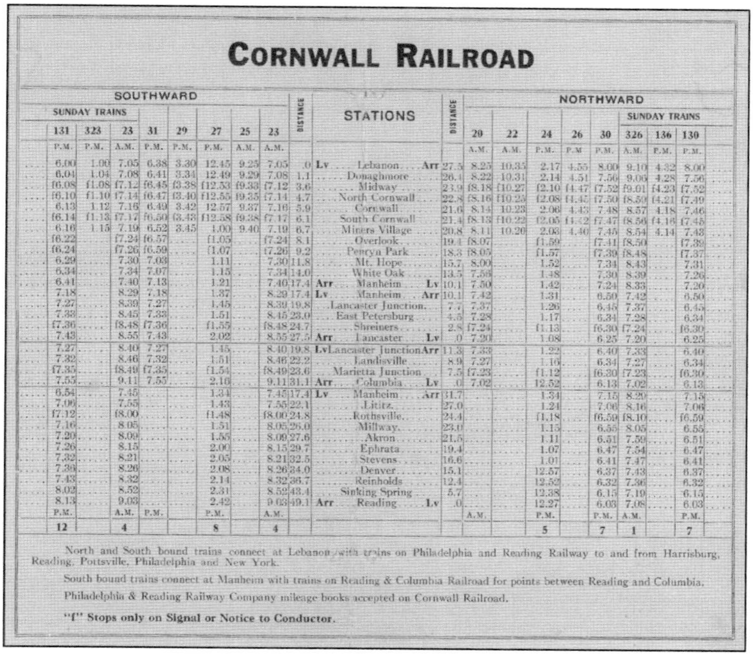

CORNWALL RAILROAD

SOUTHWARD									STATIONS		NORTHWARD							
SUNDAY TRAINS								DISTANCE		DISTANCE						SUNDAY TRAINS		
131	323	23	31	29	27	25	23				20	22	24	26	30	326	136	130
P.M.	P.M.	A.M.	P.M.	P.M.	P.M.	A.M.	A.M.				A.M.	A.M.	P.M.	P.M.	P.M.	A.M.	P.M.	P.M.
6.00	1.00	7.05	6.38	3.30	12.45	9.25	7.05	.0	Lv Lebanon Arr	27.5	8.25	10.35	2.17	4.55	8.00	9.10	4.32	8.00
6.04	1.04	7.08	6.41	3.34	12.49	9.29	7.08	1.1	Donaghmore	26.4	8.22	10.31	2.14	4.51	7.56	9.06	4.28	7.56
f6.08	f1.08	f7.12	f6.45	f3.38	f12.53	f9.33	f7.12	3.6	Midway	23.9	f8.18	f10.27	f2.10	f4.47	f7.52	f9.01	f4.23	f7.52
f6.10	f1.10	f7.14	f6.47	f3.40	f12.55	f9.35	f7.14	4.7	North Cornwall	22.8	f8.16	f10.25	f2.08	f4.45	f7.50	f8.59	f4.21	f7.49
6.13	1.12	7.16	6.49	3.42	12.57	9.37	7.16	5.9	Cornwall	21.6	8.14	10.23	2.06	4.43	7.48	8.57	4.18	7.46
f6.14	f1.13	f7.17	f6.50	f3.43	f12.58	f9.38	f7.17	6.1	South Cornwall	21.4	f8.13	f10.22	f2.05	f4.42	f7.47	f8.56	f4.16	f7.45
6.16	1.15	7.19	6.52	3.45	1.00	9.40	7.19	6.7	Miners Village	20.8	8.11	10.20	2.03	4.40	7.45	8.54	4.14	7.43
f6.22		f7.24	f6.57		f1.05		f7.24	8.1	Overlook	19.4	f8.07		f1.59		f7.41	f8.50		f7.39
f6.24		f7.26	f6.59		f1.07		f7.26	9.2	Penryn Park	18.3	f8.05		f1.57		f7.39	f8.48		f7.37
6.29		7.30	7.03		1.11		7.30	11.8	Mt. Hope	15.7	8.00		1.52		7.34	8.43		7.31
6.34		7.34	7.07		1.15		7.34	14.0	White Oak	13.5	7.56		1.48		7.30	8.39		7.26
6.41		7.40	7.13		1.21		7.40	17.4	Arr Manheim Lv	10.1	7.50		1.42		7.24	8.33		7.20
7.18		8.29	7.18		1.37		8.29	17.4	Lv Manheim Arr	10.1	7.42		1.31		6.50	7.42		6.50
7.27		8.39	7.27		1.45		8.39	19.8	Lancaster Junction	7.7	7.37		1.26		6.45	7.37		6.45
7.33		8.45	7.33		1.51		8.45	23.0	East Petersburg	4.5	7.28		1.17		6.34	7.28		6.34
f7.36		f8.48	f7.36		f1.55		f8.48	24.7	Shreiners	2.8	f7.24		f1.13		f6.30	f7.24		f6.30
7.43		8.55	7.43		2.02		8.55	27.5	Arr Lancaster Lv	.0	7.20		1.08		6.25	7.20		6.25
7.27		8.40	7.27		1.45		8.40	19.8	Lv Lancaster Junction Arr	11.3	7.33		1.22		6.40	7.33		6.40
7.32		8.46	7.32		1.51		8.46	22.2	Landisville	8.9	7.27		1.16		6.34	7.27		6.34
f7.35		f8.49	f7.35		f1.54		f8.49	23.6	Marietta Junction	7.5	f7.23		f1.12		f6.30	f7.23		f6.30
7.55		9.11	7.55		2.16		9.11	31.1	Arr Columbia Lv	.0	7.02		12.52		6.13	7.02		6.13
6.54		7.45			1.34		7.45	17.4	Lv Manheim Arr	31.7			1.34		7.15	8.29		7.15
7.06		7.55			1.43		7.55	22.1	Lititz	27.0			1.24		7.06	8.16		7.06
f7.12		f8.00			f1.48		f8.00	24.8	Rothsville	24.4			f1.18		f6.59	f8.10		f6.59
7.16		8.05			1.51		8.05	26.0	Millway	23.0			1.15		6.55	8.05		6.55
7.20		8.09			1.55		8.09	27.6	Akron	21.5			1.11		6.51	7.59		6.51
7.26		8.15			2.00		8.15	29.7	Ephrata	19.4			1.07		6.47	7.54		6.47
7.32		8.21			2.05		8.21	32.5	Stevens	16.6			1.01		6.41	7.47		6.41
7.36		8.26			2.08		8.26	34.0	Denver	15.1			12.57		6.37	7.43		6.37
7.43		8.32			2.14		8.32	36.7	Reinholds	12.4			12.52		6.32	7.36		6.32
8.02		8.52			2.31		8.52	43.4	Sinking Spring	5.7			12.38		6.15	7.19		6.15
8.13		9.03			2.42		9.03	49.1	Arr Reading Lv	.0			12.27		6.03	7.08		6.03
P.M.		A.M.	P.M.		P.M.		A.M.				A.M.		P.M.		P.M.	A.M.		P.M.
12		4			8		4						5		7	1		7

North and South bound trains connect at Lebanon with trains on Philadelphia and Reading Railway to and from Harrisburg, Reading, Pottsville, Philadelphia and New York.

South bound trains connect at Manheim with trains on Reading & Columbia Railroad for points between Reading and Columbia.

Philadelphia & Reading Railway Company mileage books accepted on Cornwall Railroad.

"f" Stops only on Signal or Notice to Conductor.

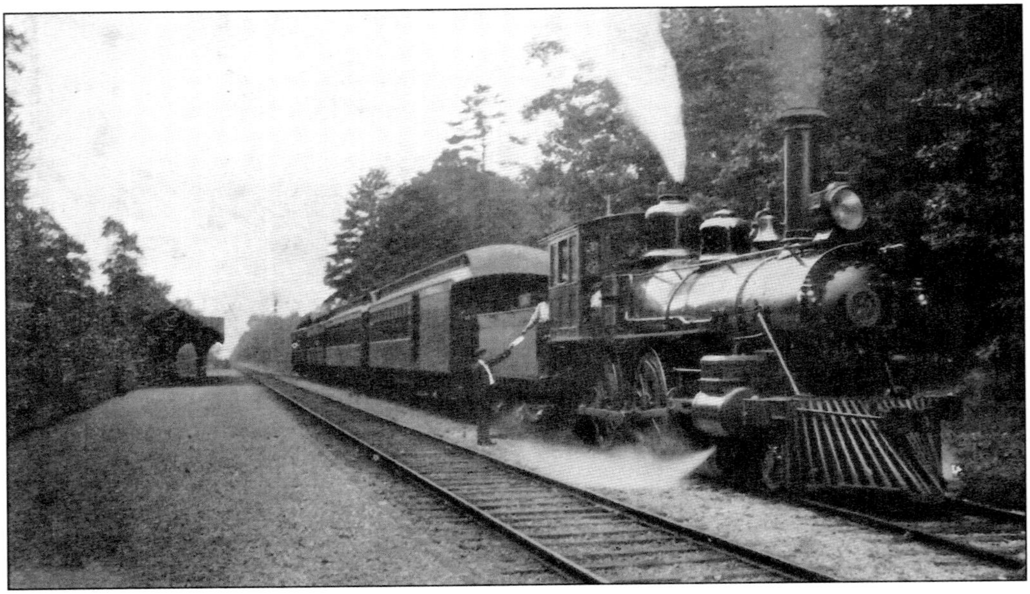

A Cornwall Railroad passenger train arrives at Penryn Park (above). The station can be seen on the left. The park included a lake for swimming and boating, a merry-go-round, a whip, a baseball field with bleachers, a large pavilion for dancing, and picnic tables under the trees, which can be seen in the other photograph (below). The park was built in the mid-1880s and was a favorite recreation destination for families from Cornwall and surrounding towns. It was also popular for church picnics, among other things. For many years, this facility was operated as the Lebanon YMCA's Camp Kiwanis. Currently, the Lancaster YMCA operates the park as Camp Shand. (Both, courtesy of Daniel Chernich.)

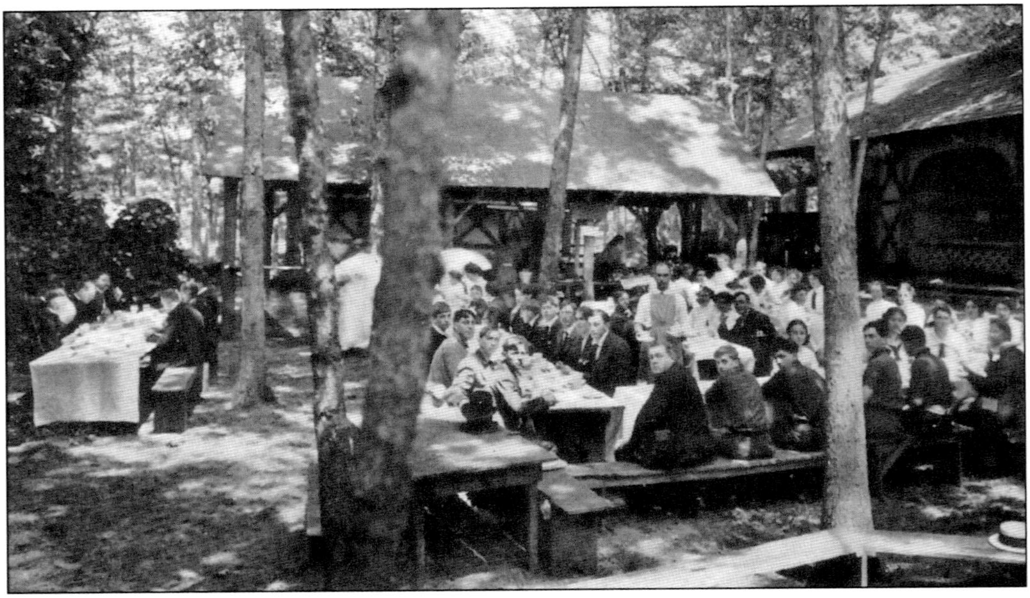

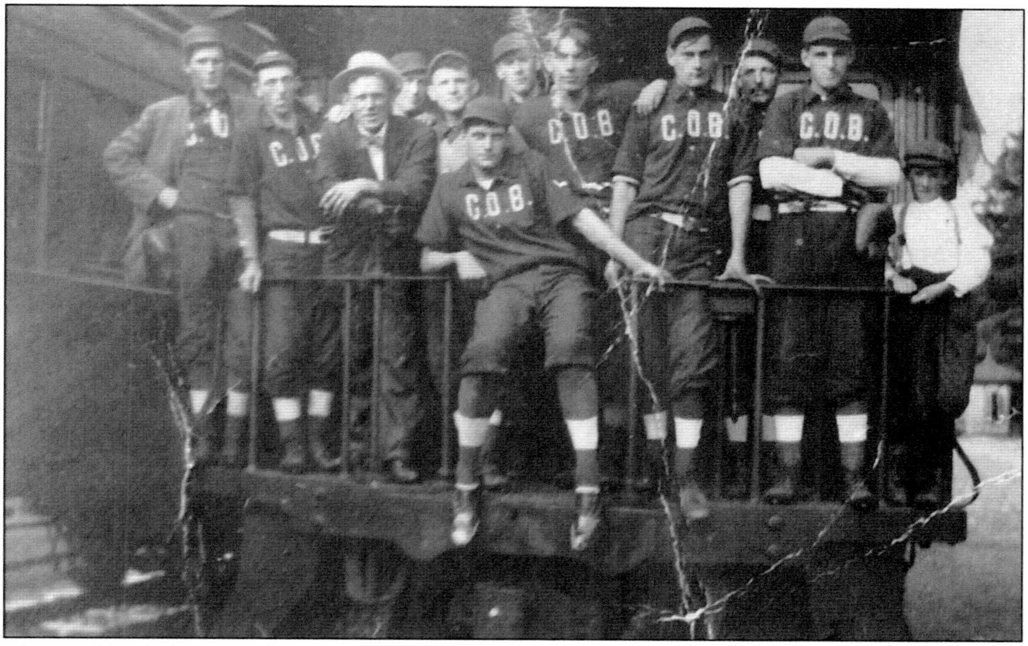

This undated photograph shows the Cornwall Ore Bank Company baseball team traveling on the Cornwall Railroad.

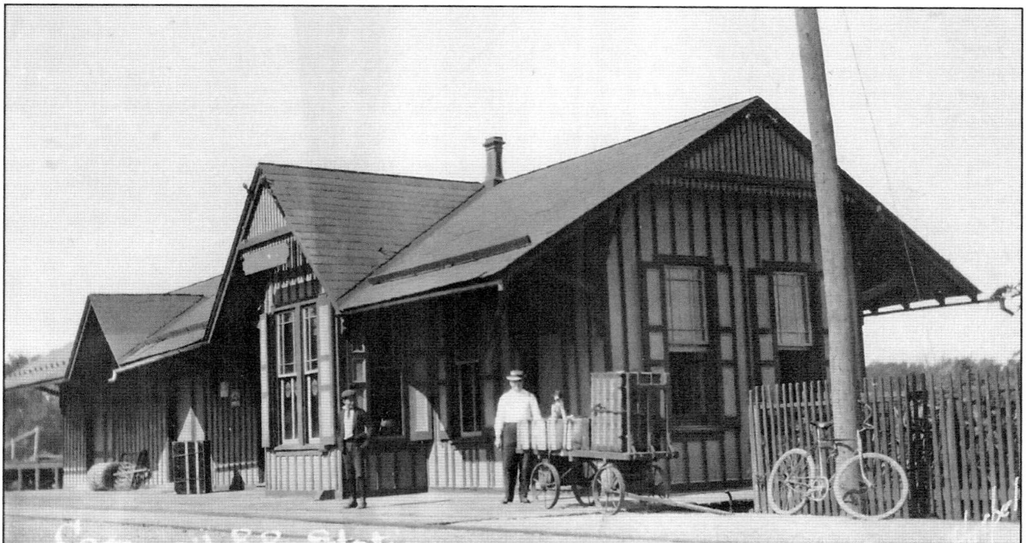

The Cornwall Railroad Station operated as long as passenger and freight services were offered to Lebanon or other locations. After transporting over five million passengers, the railroad ended passenger service ended in 1931. Freight service continued until the early 1950s. The Borough of Cornwall bought the building and remodeled and expanded it. Today, the borough uses the structure as a police station and for administrative offices. (Courtesy of Daniel Chernich.)

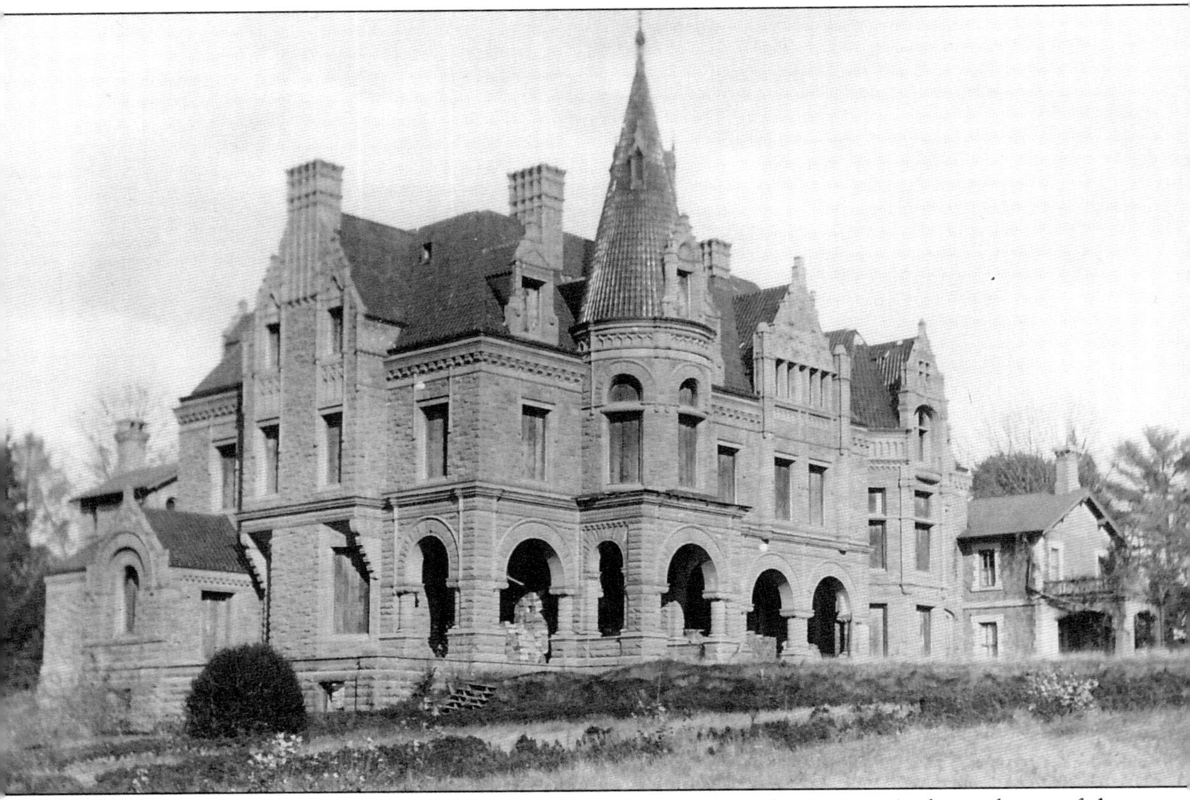

This photograph shows Robert H. Coleman's nearly completed mansion. A close relative of the owners of the Cornwall Railroad, Coleman built his own competing rail line. His new railway was known as the Cornwall & Lebanon Railroad (C&L). Even though it had no connection with the Cornwall Railroad, its tracks ran parallel with the Cornwall's rails. After his young wife died, Coleman had the mansion torn down. The stone was taken to Lebanon and used to build St. Luke's Episcopal Church. Nothing remains of the home in Cornwall. In 1918, the Pennsylvania Railroad acquired the Cornwall & Lebanon Railroad. In 1999, the Rails-to-Trails Conservancy converted the C&L rail bed to a public bike and walking trail.

Four

Bethlehem Steel and Cornwall

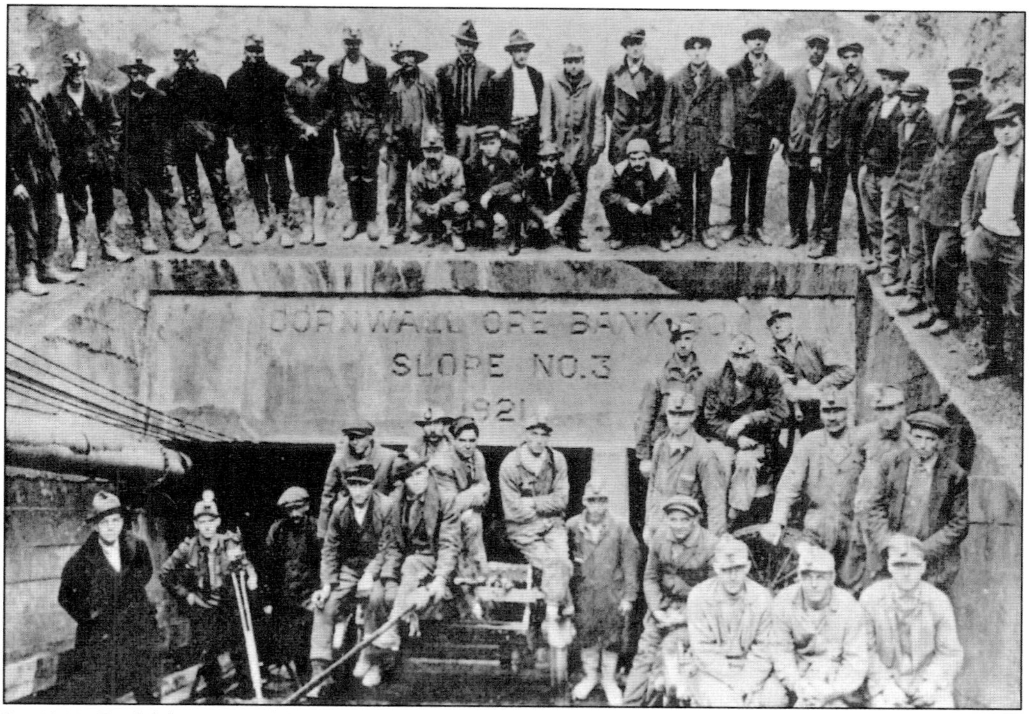

In 1919, Bethlehem Steel became the majority owner of the mines at Cornwall and, in 1926, purchased the mining rights of the Robesonia Iron Company and became the sole owner of that mine. In 1921, the mining operation went underground. A shaft was sunk in the footwall at the west end of the open pit so that the lower extremities of the original deposit could be recovered. New hires for underground labor were needed. Enticed by the prospect of a job that offered decent pay and boarding in well-built stone homes with room for a garden, workers came from as far away as Minnesota. Immigrants arrived from Italy, Germany, England, Wales, Hungary, Czechoslovakia, and Mexico. For many years, up to 16 different languages were spoken inside the mines.

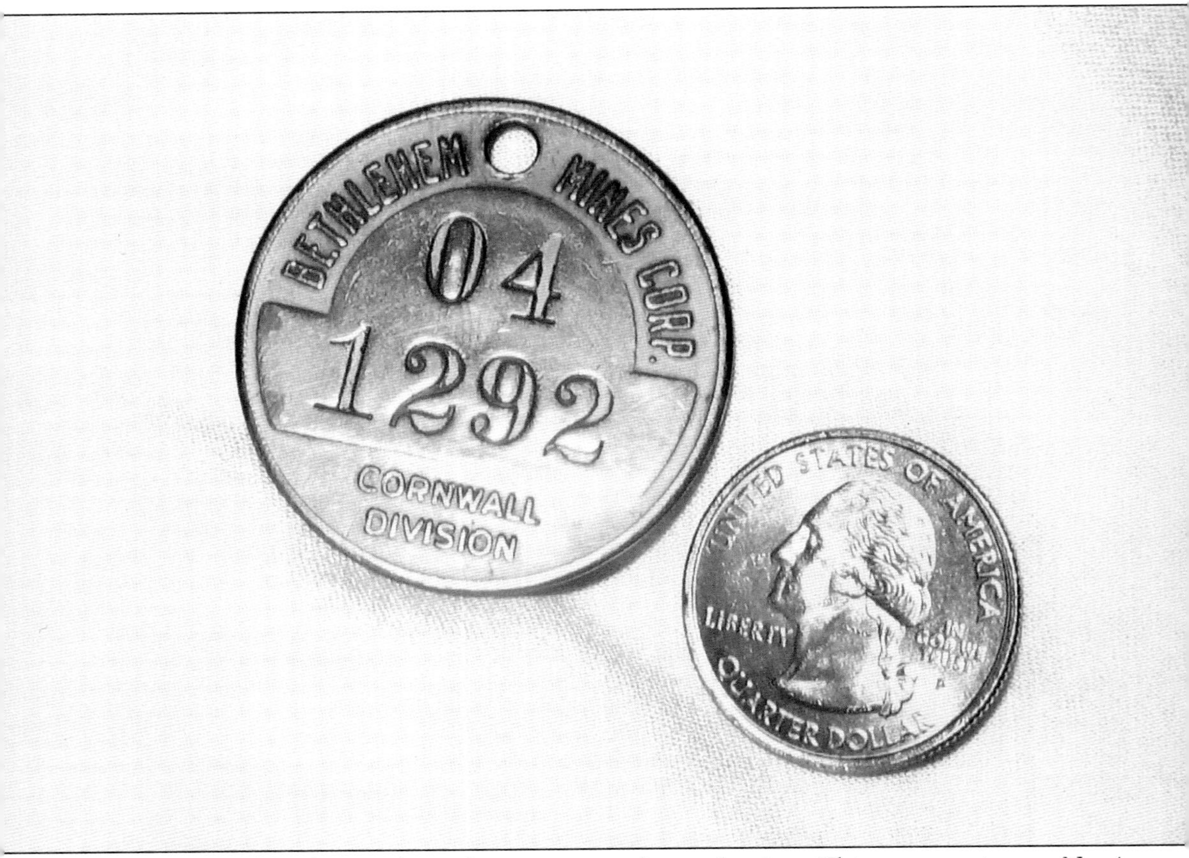

Brass checks, like the one shown here, were issued to each miner. This one was given to No. 4 Mine workers. Those issued at No. 3 Mine would have included a *03* above a three-digit number but were otherwise identical to this one. Many mines required such checks be hung on a board or another fixture at the beginning of the shift and removed at the end of the shift so management could see who was in the mine. Miners at Cornwall were not required to display their checks during their work hours; the brass coins were used only to receive pay.

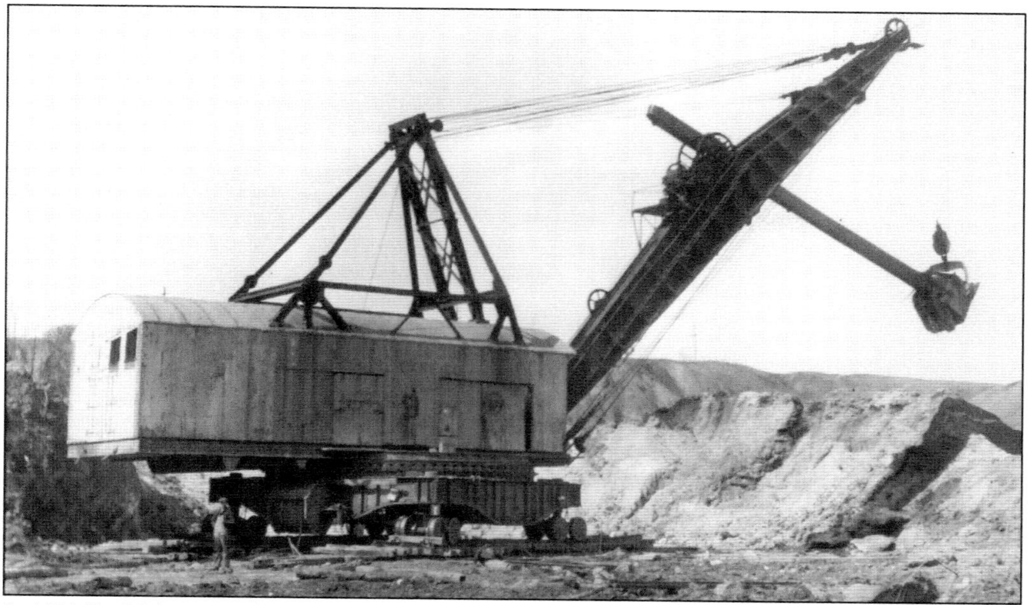

In 1921, Bethlehem Steel bought an all-electric Marion shovel with a six-cubic-yard dipper. Because the shovel's dipper was so large, Bethlehem Steel had to purchase more drills and increase the size of its railcars. Whereas the old steam shovels needed only two drills to make holes for blasting ahead of them, the electric shovel required six drills ahead of it to supply enough stock for the shovel to work continuously.

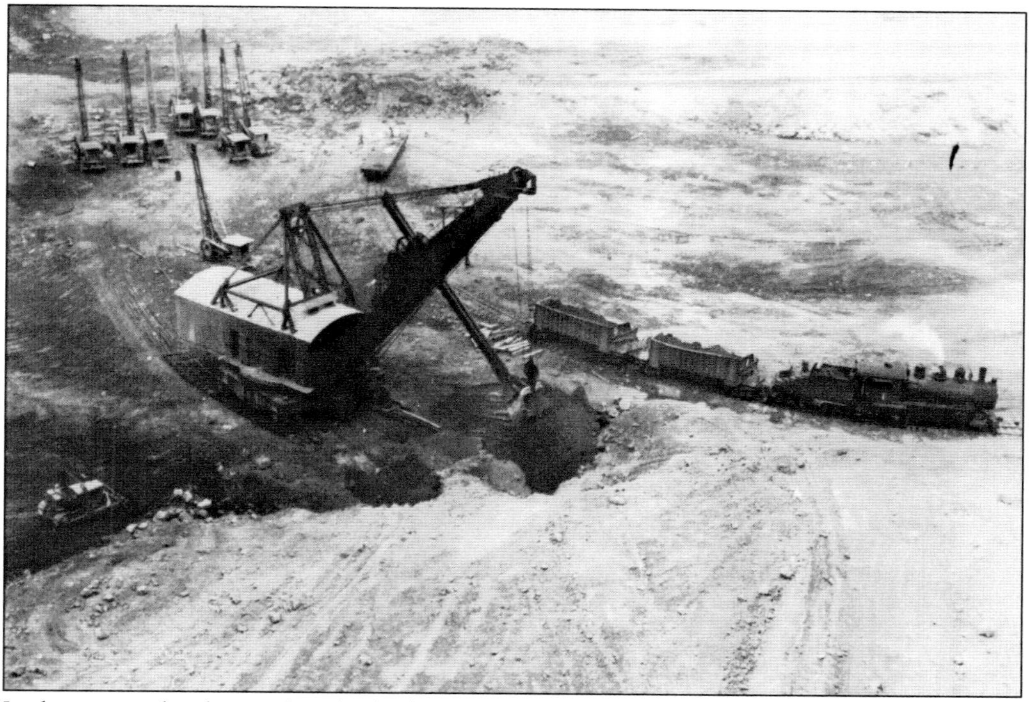

In this image, the electric shovel is loading railcars. Around the same time the electric shovel was purchased, the track gear on the old steam shovels was removed and replaced with Caterpillar bulldozer treads. With this new and retooled equipment, the mines, operating on two eight-hour shifts each day, removed five million tons of rock and produced one million tons of ore annually.

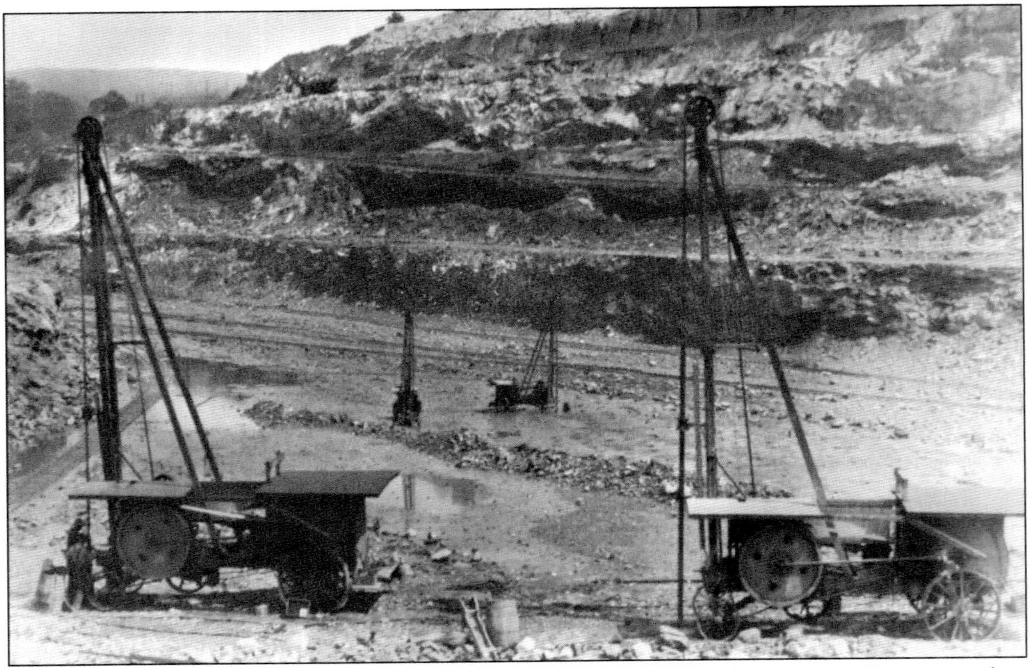

These new well-drilling rigs replaced the old steam drills. Early gasoline engines known as "hit-and-miss" or "make-or-break" engines powered the new drills, which were used to create holes in the rock. The holes were then loaded with dynamite and blasted. A blast can be seen in the picture below. The drills were needed to keep loose material ahead of the shovels.

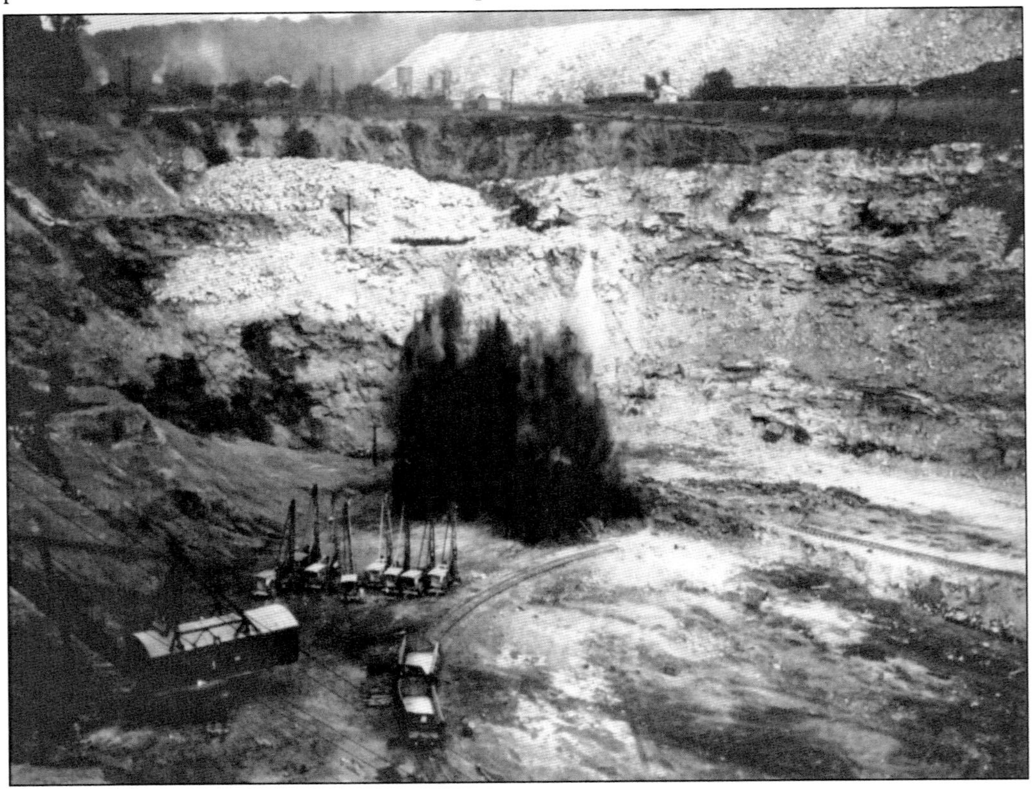

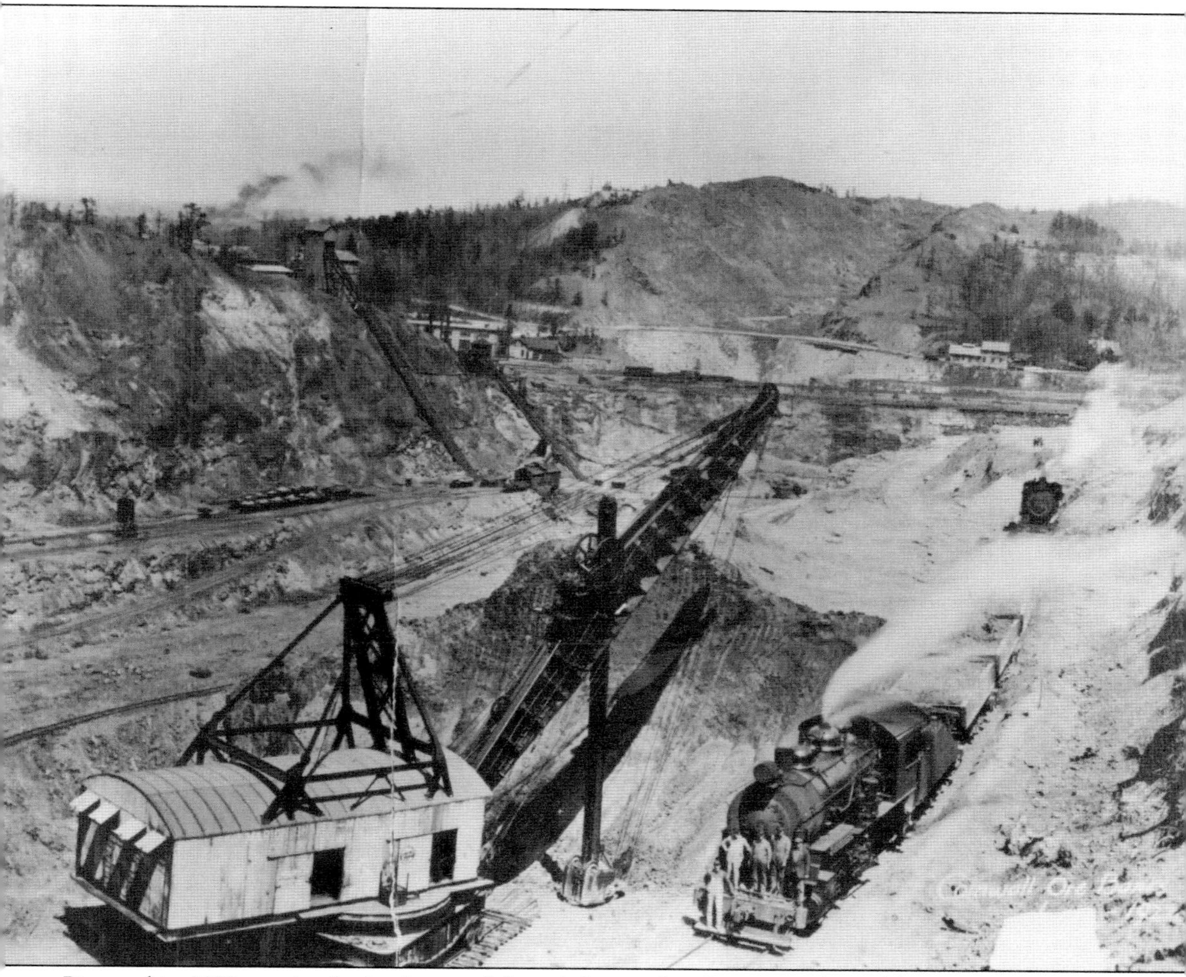

Pictured in 1922 are, in the background from left to right, the two inclines (directly above the steam shovel), the original open-pit operation on Big Hill, and the Cornwall Ore Bank office. The taller incline belongs to the Cornwall Ore Bank Company and the shorter one to the Robesonia Iron Company. The strip of land directly behind the top of the boom divides the open pit and supports tracks for the railroad that services the Robesonia headframe and hauls away ore. In this image, a portion of the dividing wall has been removed. In later years, the wall was completely removed, creating one large open-pit mine. Railroad tracks now run under the bridge where part of the rock wall was taken away.

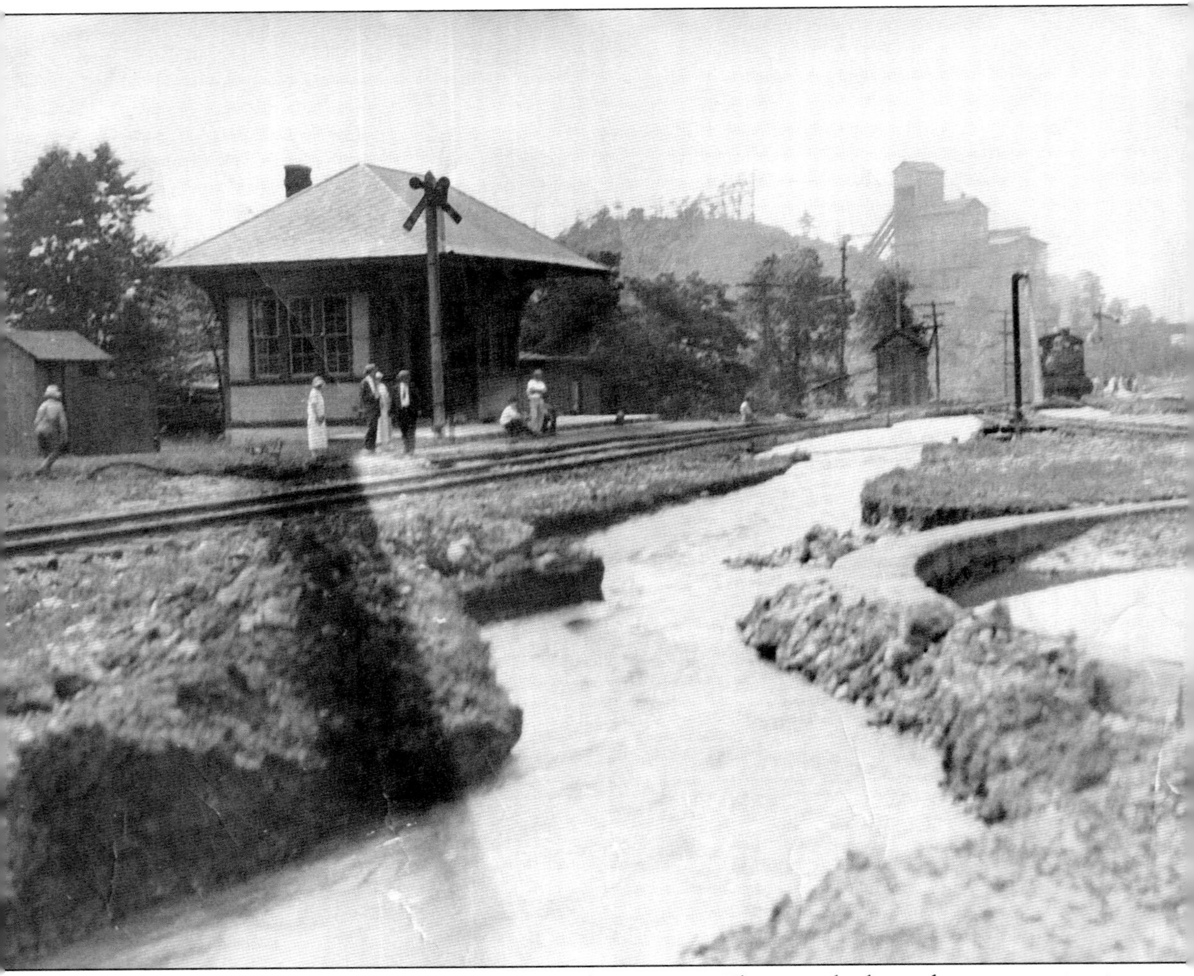

In July 1925, heavy rains caused flooding in the open pit. This view looks at the train station in Minersvillage, with the Cornwall Railroad tracks in front. The stone arc on the right side of the photograph is part of the turntable for the Cornwall Ore Bank Company engines. The people shown in the photograph have ventured out to watch the floodwaters rush into the open pit after the rain stopped. In the background, the headframe of the No. 1 Incline can be seen.

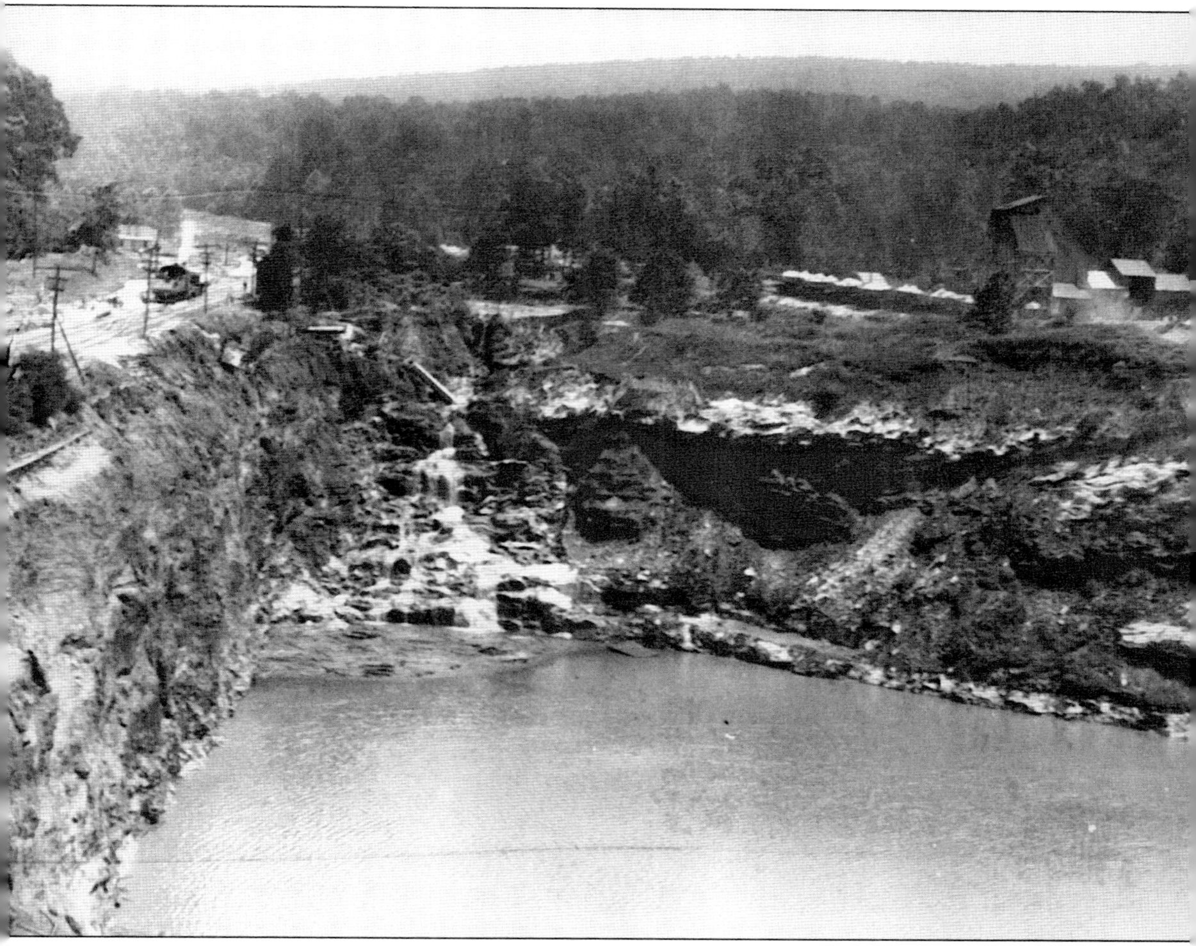

Remaining floodwater pours into the open pit on its east side. The mountains behind Minersvillage and the lay of the land allowed floodwater to follow the natural flow of the creeks and into the pit. Amazingly, the floodwater from Hurricane Agnes would travel the same way some 67 years later.

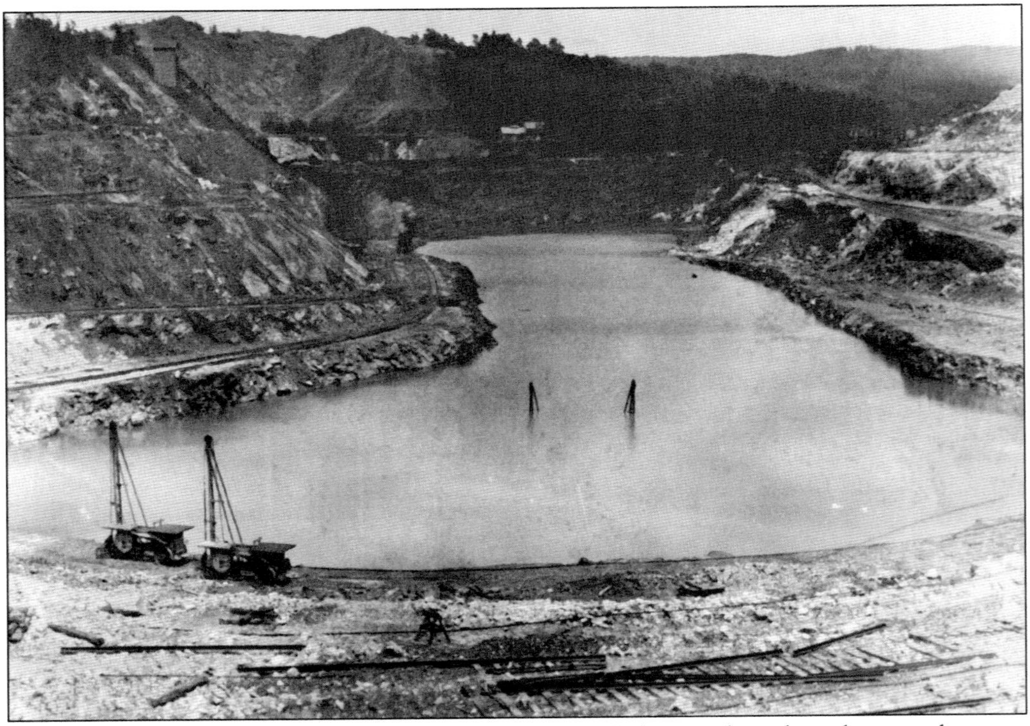

In these photographs, floodwater stands in the open pit. Indicating how deep the water became, just the tops of the well drills are sticking out over the surface.

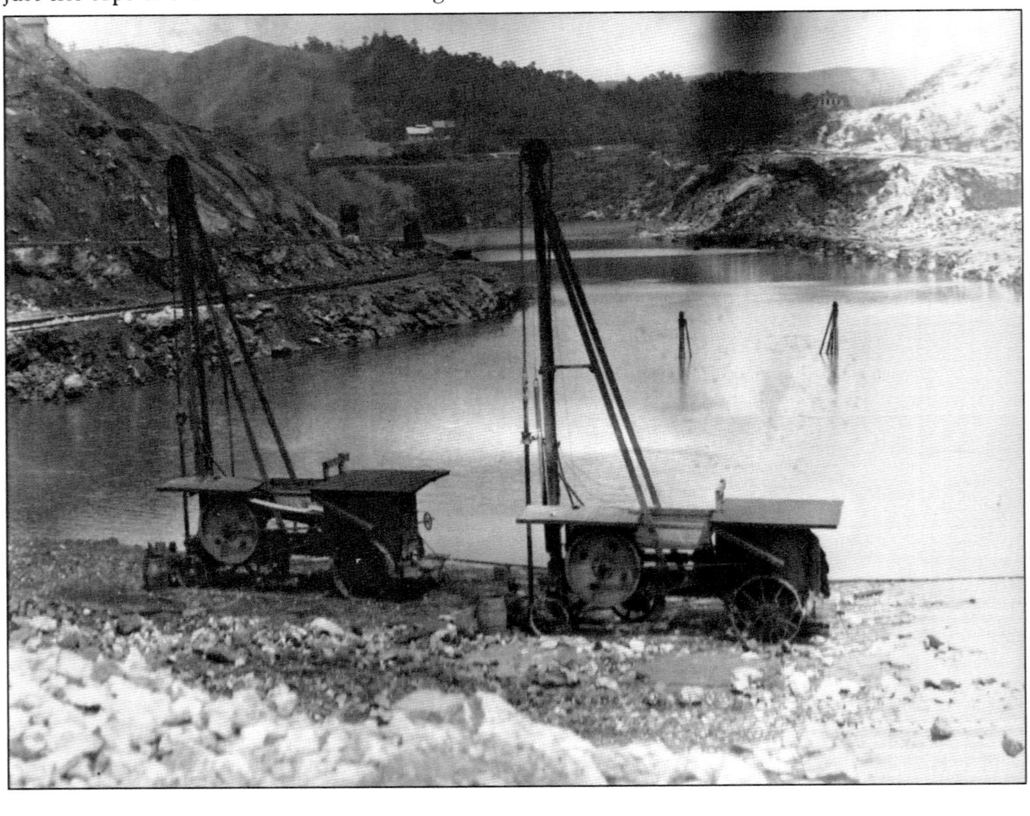

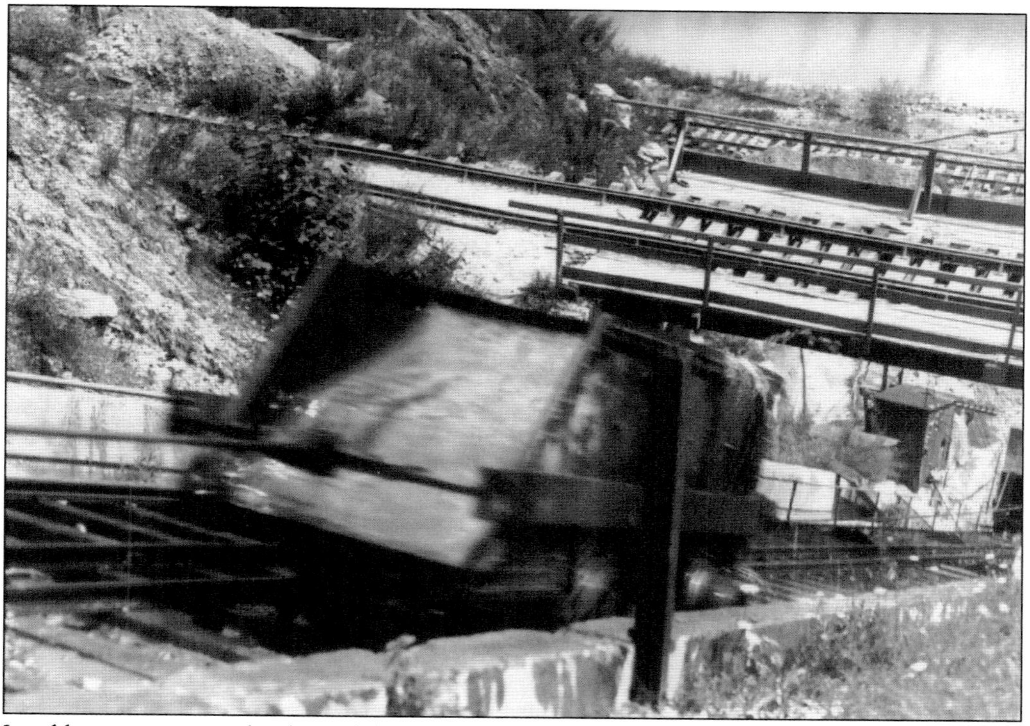

In addition to pumps, the skips of the No. 1 Incline were used to lift water from the mine. These pictures show the skips bringing water up to be dumped.

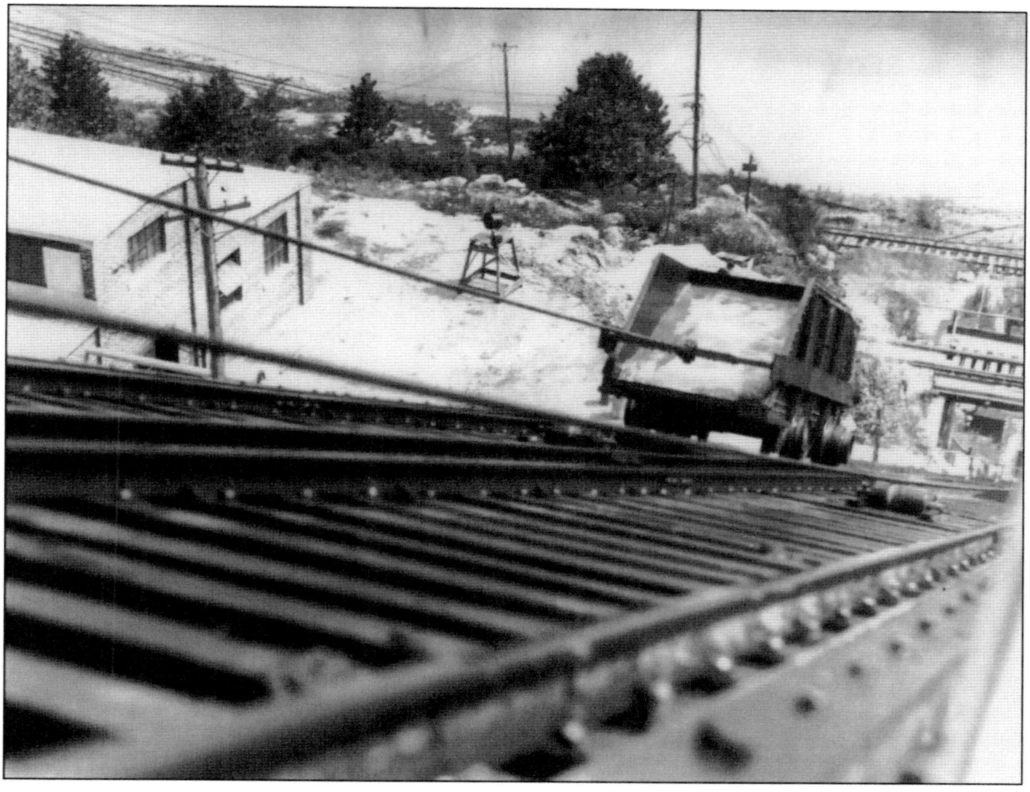

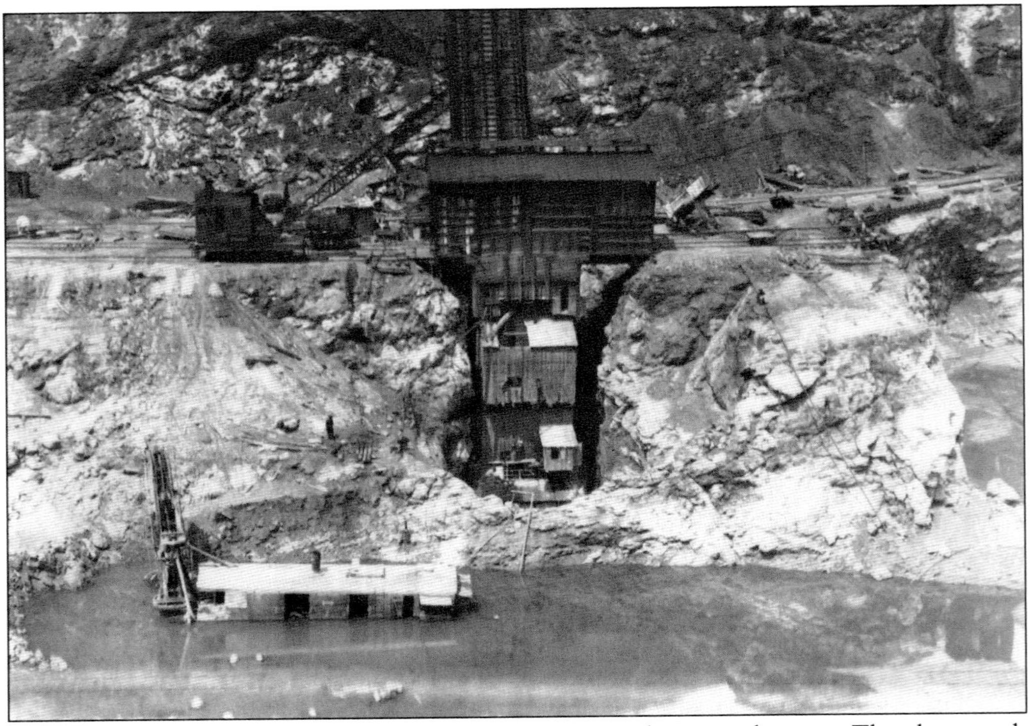

In the image above, the steam shovel next to the No. 1 Incline is underwater. The photograph below shows miners working to dig ore, mud, and flood debris away from another steam shovel.

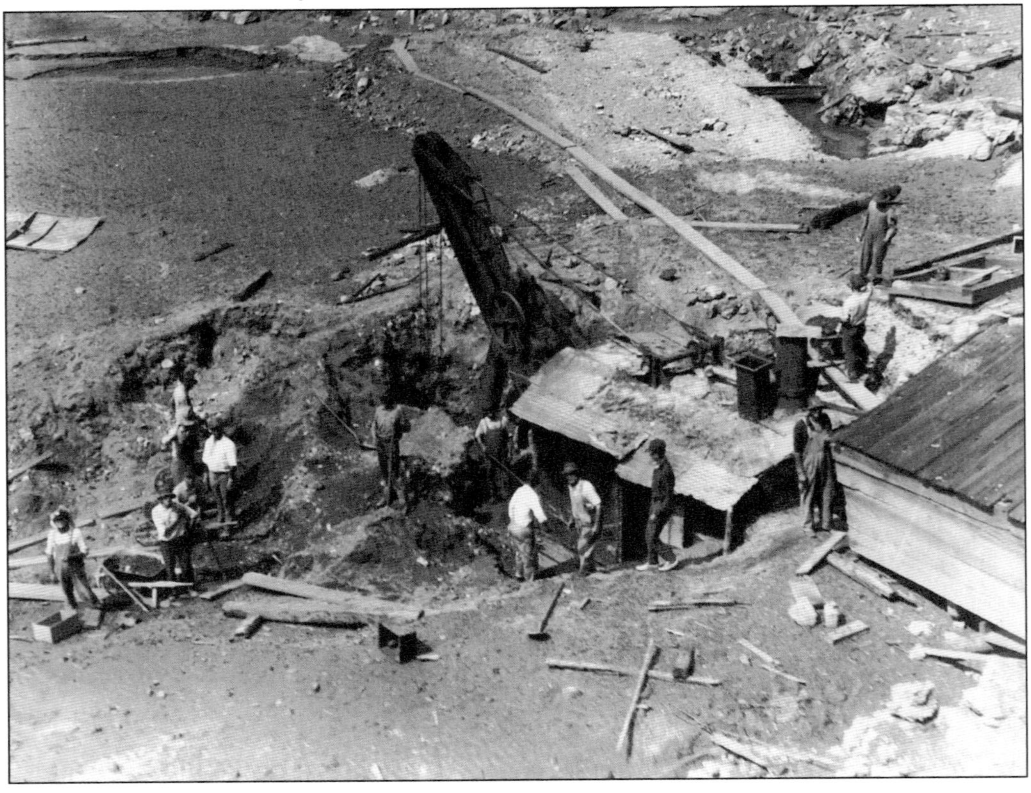

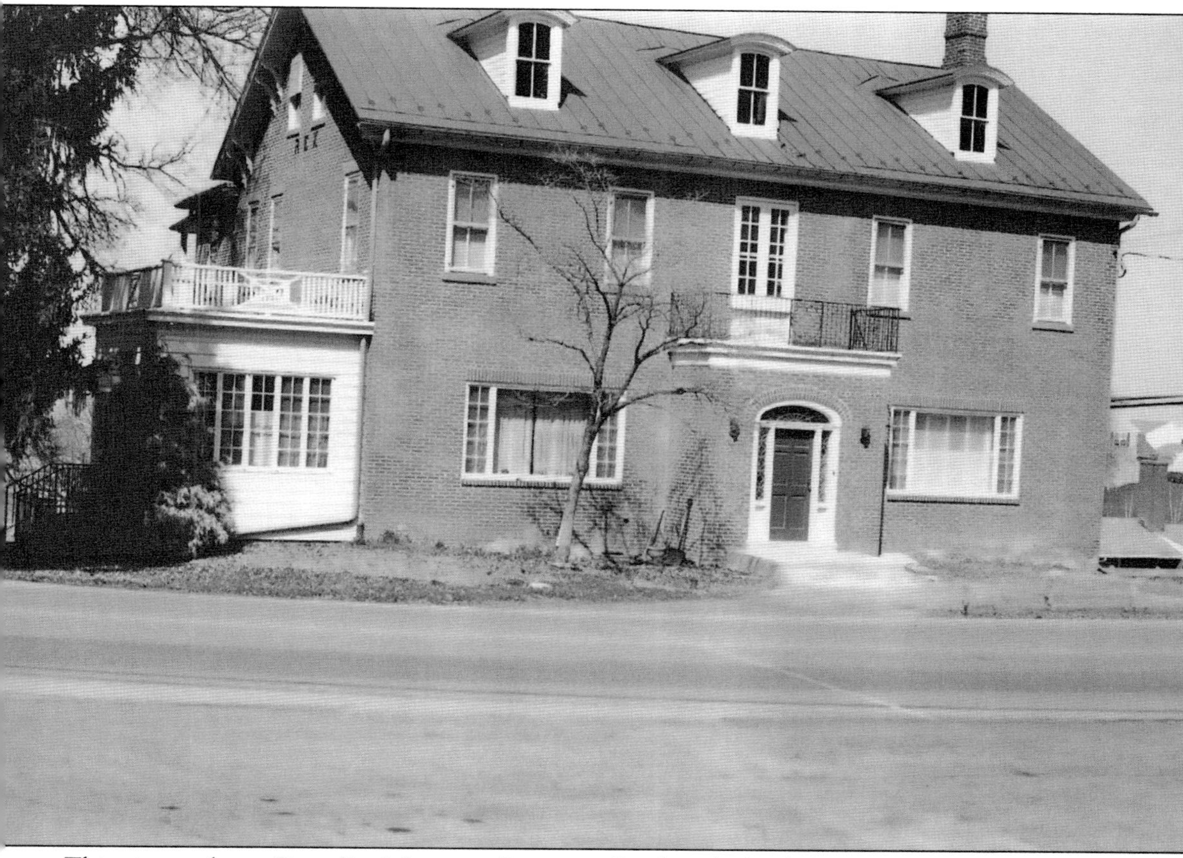

This picture shows Cyrus Rex's house in Rexmont. Rex bought land cleared by the woodcutters and colliers and divided the land into lots for homes in what would become Rexmont. He also sold cleared land for farms in the area around the community. People who wished to own their homes rather than pay rent to the mining companies built the houses in Rexmont.

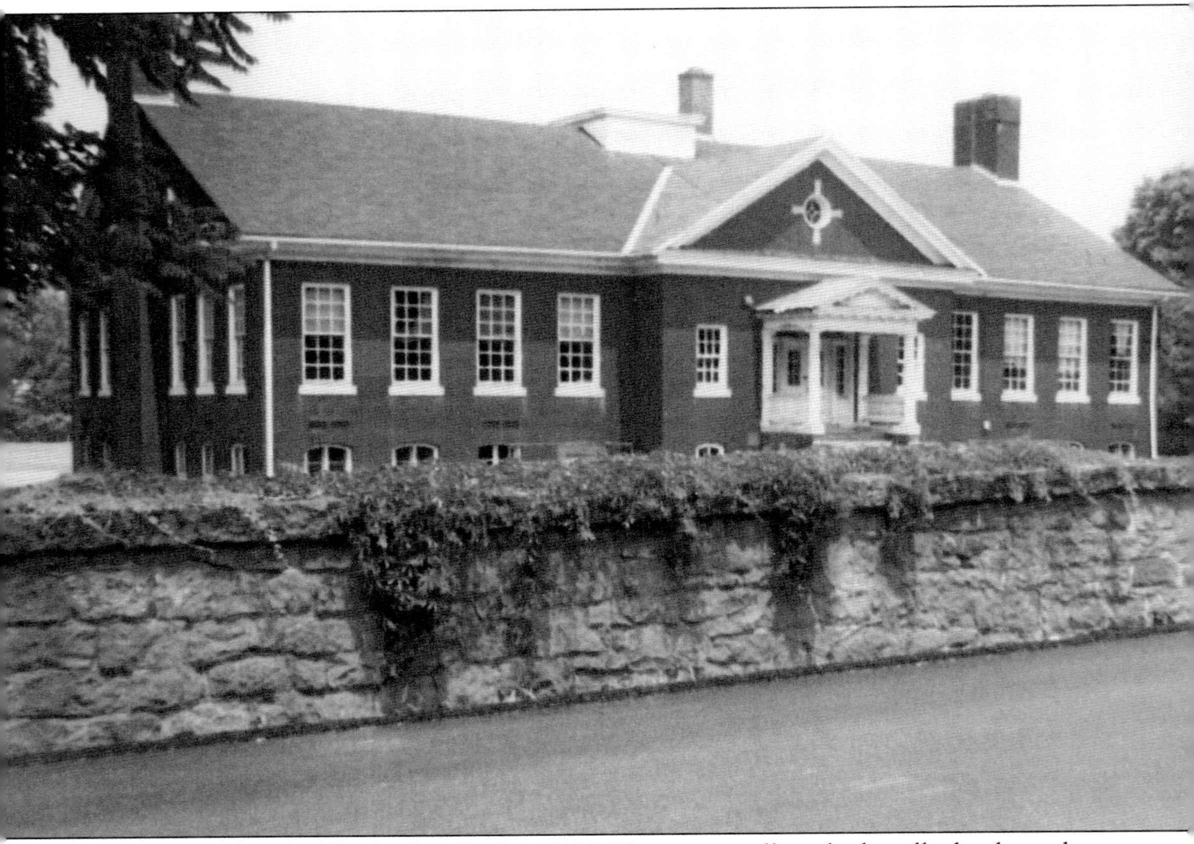

This is Cornwall's old high school built in 1902. The mining villages had small schools to educate the children until they reached high school age. When they were old enough, these students transferred to Cornwall's high school. At one time, a dome sat atop the high school building, but it had been removed by the time this photograph was taken. The entire building was razed in 2008. (Courtesy of Daniel Chernich.)

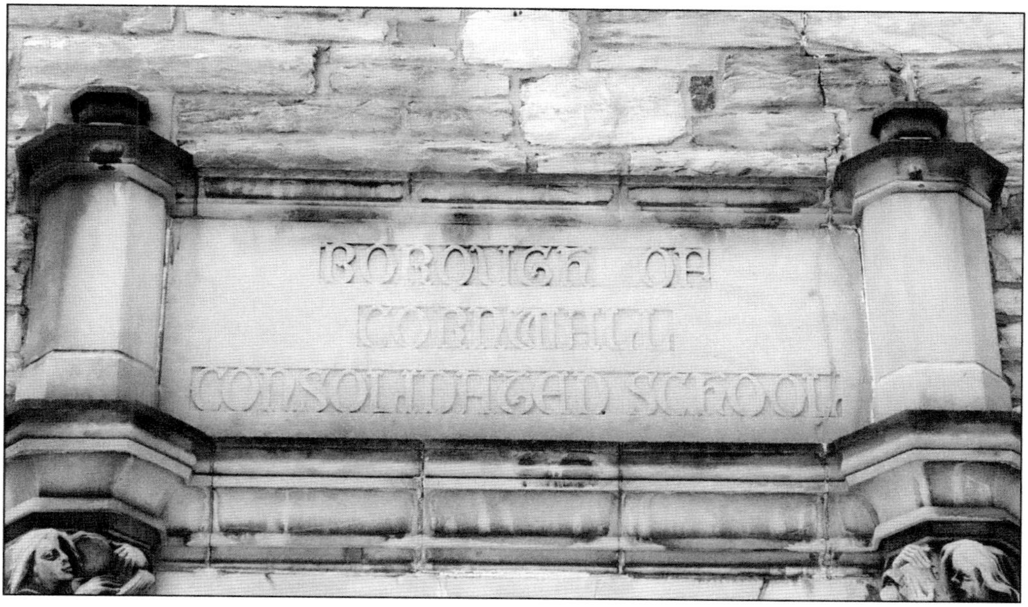

In 1927, Cornwall built a new school. An inscription in the stonework over the main doorway reads, "Borough of Cornwall Consolidated School." When this building was completed in 1928, the small village schools were closed. The consolidated school educated children from grades 1 through 12. Prof. Ray Light was the principal, making it the only school in Lebanon County with an actual college professor as chief administrator. Many of the miners were immigrants, and most had limited educations. The community was proud to have such a nice school to give the children an opportunity for an education. The picture below shows the entire front of the building.

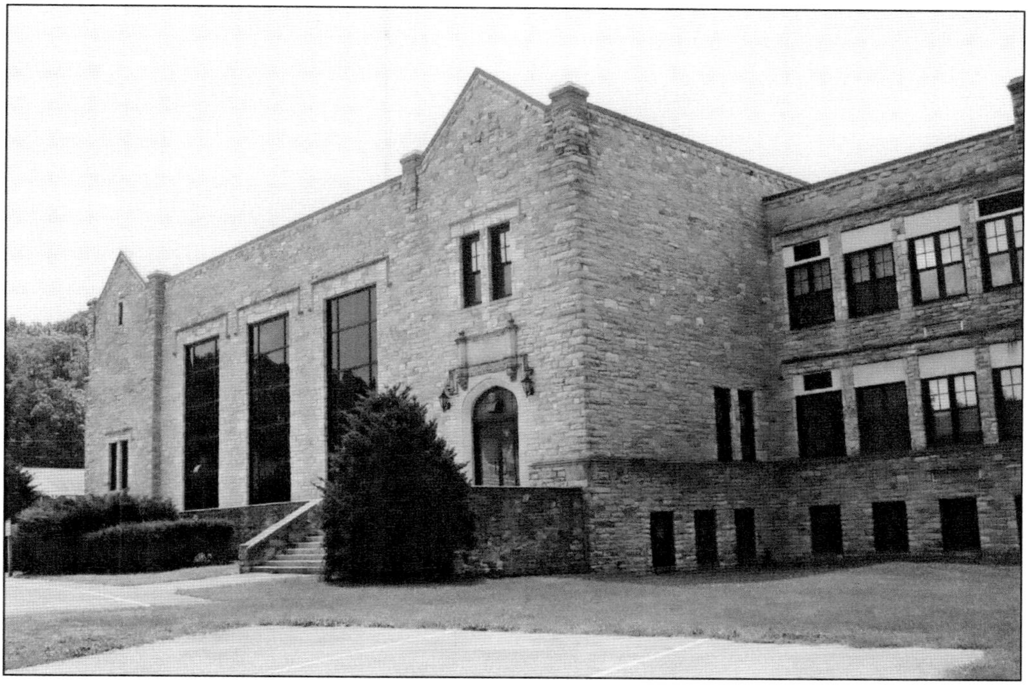

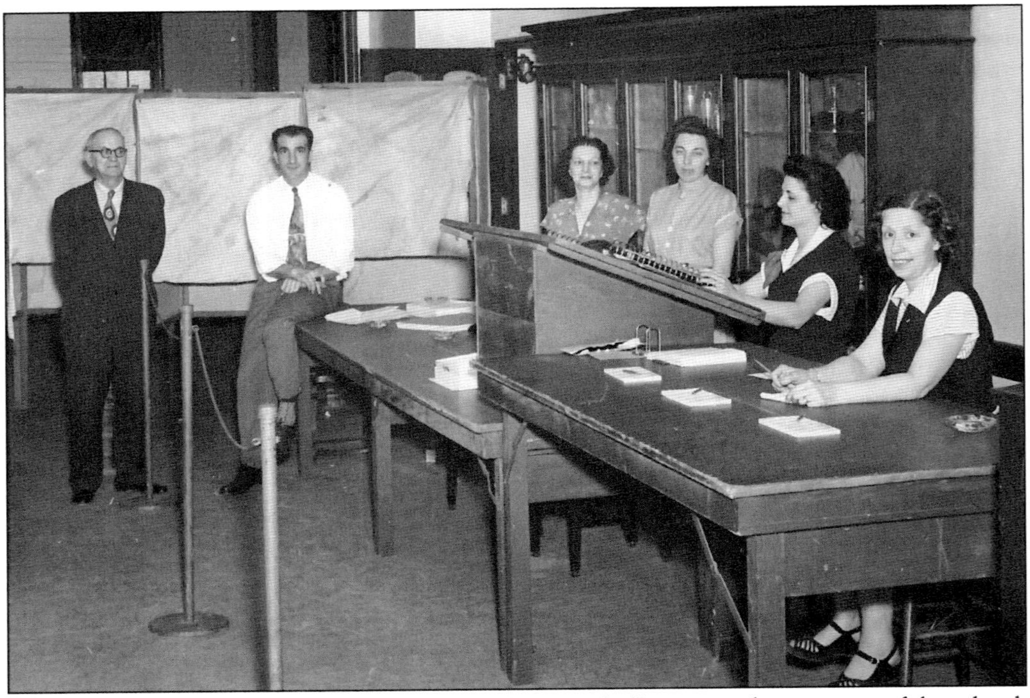

Voting for primary and general elections took place in a hallway near the entrance of the school. This photograph shows the board of election ready for the public to vote. All in Cornwall who voted did so at the consolidated school.

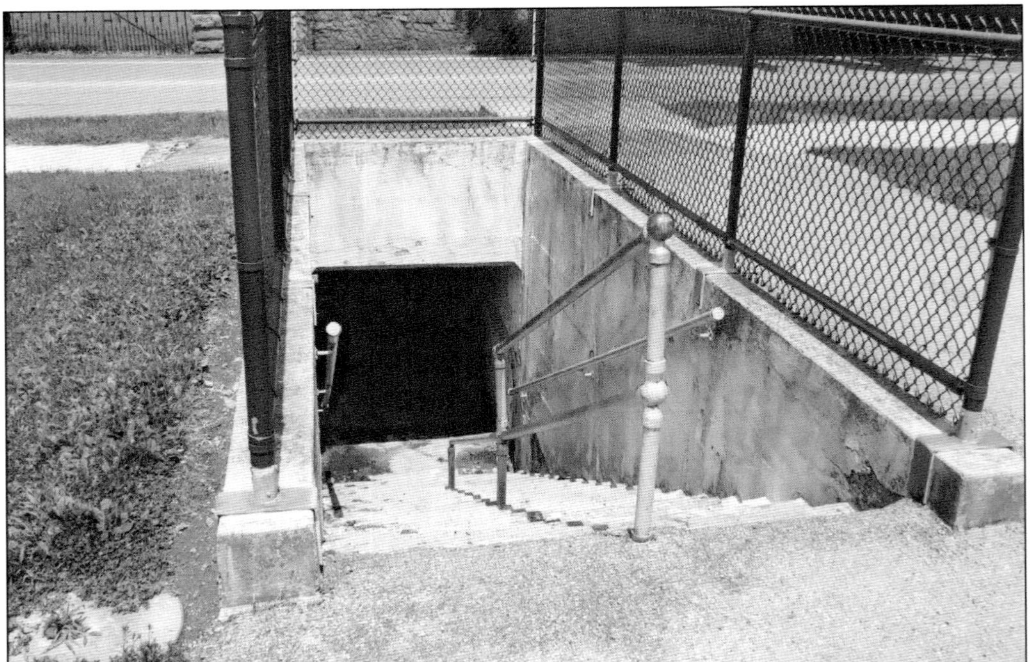

A tunnel under the road connected the new consolidated school with the older high school building. These steps lead down to the tunnel. Because some classes were held in the old building, students would traverse the tunnel so they could make it to class without having to cross the street.

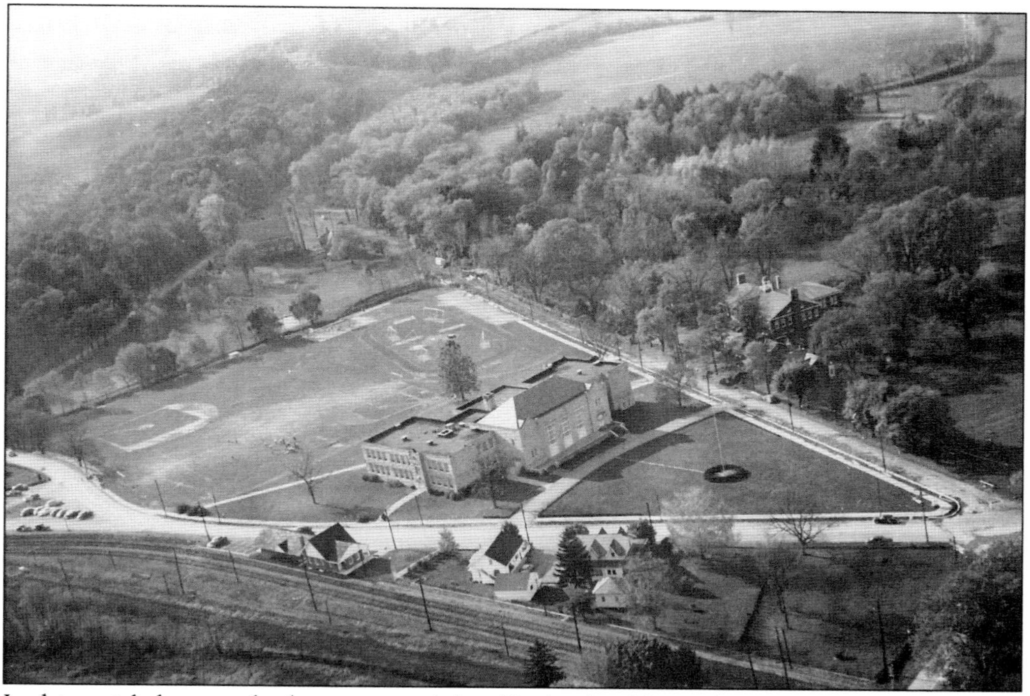

In this aerial photograph, the new consolidated high school (center) is visible along with the old brick high school building (right). In front of the school is the flagpole, and behind it are ball fields and a playground. Cornwall's train station and railroad tracks are near the bottom of the photograph. Residents of this small, isolated community worked together, knew their neighbors well, and looked out for each other.

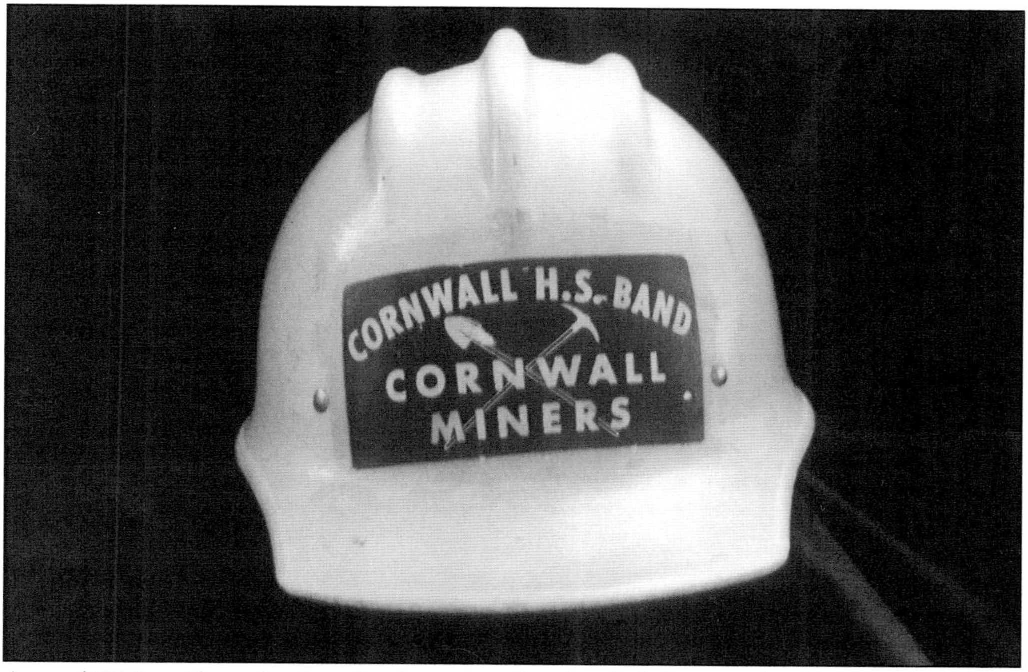

Around 1960, uniforms for the school marching band included miner's hard hats. Wherever the band performed, members wore hats such as this one.

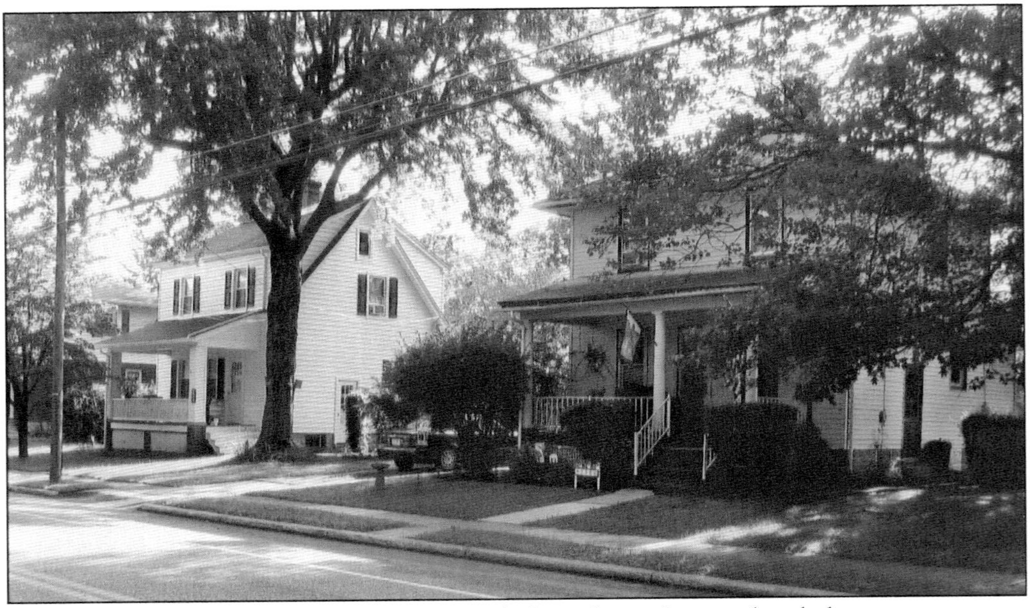

In 1927, development on No. 4 Mine began with the sinking of two inclined, three-compartment shafts in the footwall beneath the ore body. New workers were needed to manage No. 3 Mine, the open pit, and the soon-to-be-developed No. 4 Mine. Between 1929 and 1930, Bethlehem Steel built a village to provide more homes for engineers, clerks, and general office workers. The residences were built along the Cornwall Pike (on the path of the old plank road) and Pine Street. The houses were kit homes; many of the sections were put together off-site, and the structures were built by assembling the prefabricated parts. Because the community thought the homes were assembled like toys, the village was known as Toy Town. Pictured above and below are examples of Toy Town homes.

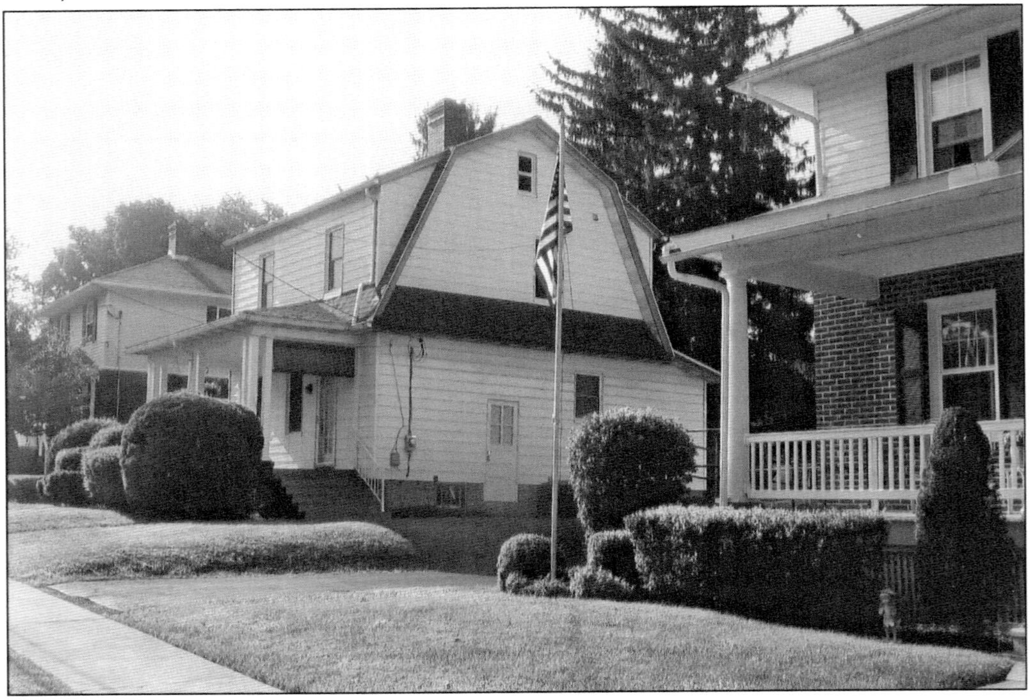

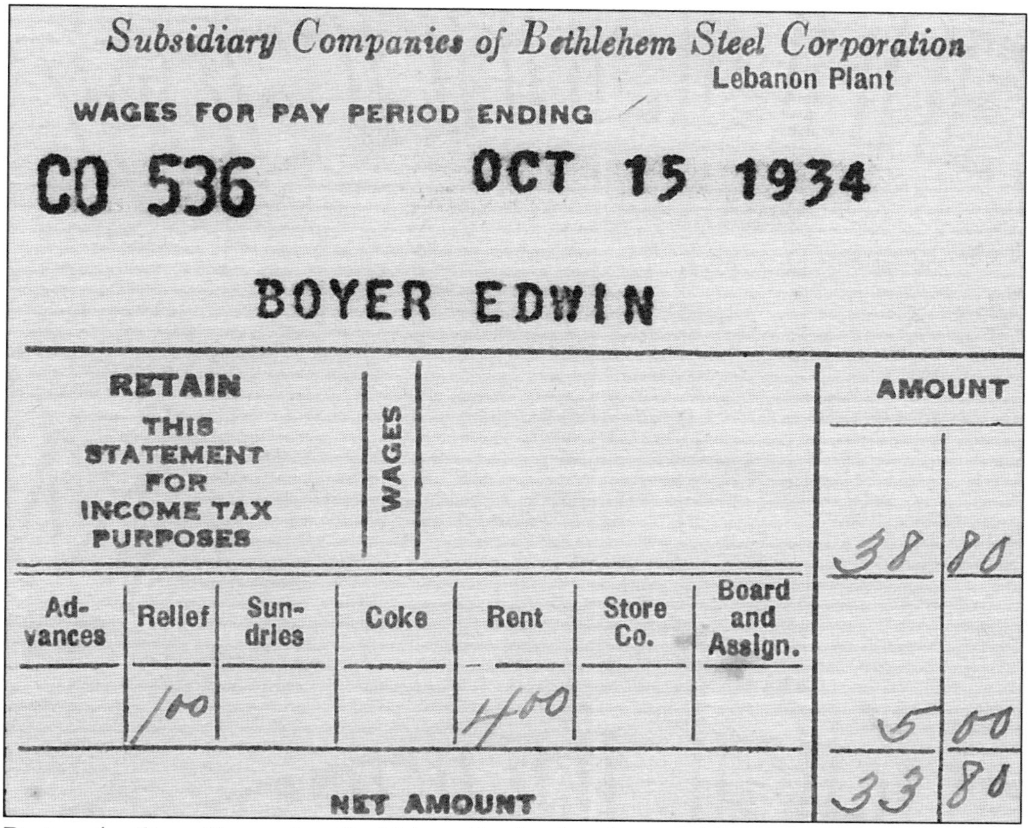

Subsidiary Companies of Bethlehem Steel Corporation
Lebanon Plant

WAGES FOR PAY PERIOD ENDING

CO 536 OCT 15 1934

BOYER EDWIN

RETAIN THIS STATEMENT FOR INCOME TAX PURPOSES

WAGES		AMOUNT
		38 80

Advances	Relief	Sundries	Coke	Rent	Store Co.	Board and Assign.	
	100			4 00			5 00

NET AMOUNT 33 80

During the Great Depression, Bethlehem Steel reduced its workforce. In 1931, the company halted development of No. 4 Mine, which was allowed to fill with water. Single men without families were laid off. Married men with children were still employed; however, they may only have worked every other day every other week. All the men on the payroll at this time were living in company houses. Most homes rented for $4 a month, but the new Toy Town residences rented for $8 a month. A pay stub from 1934 shows the rent deduction.

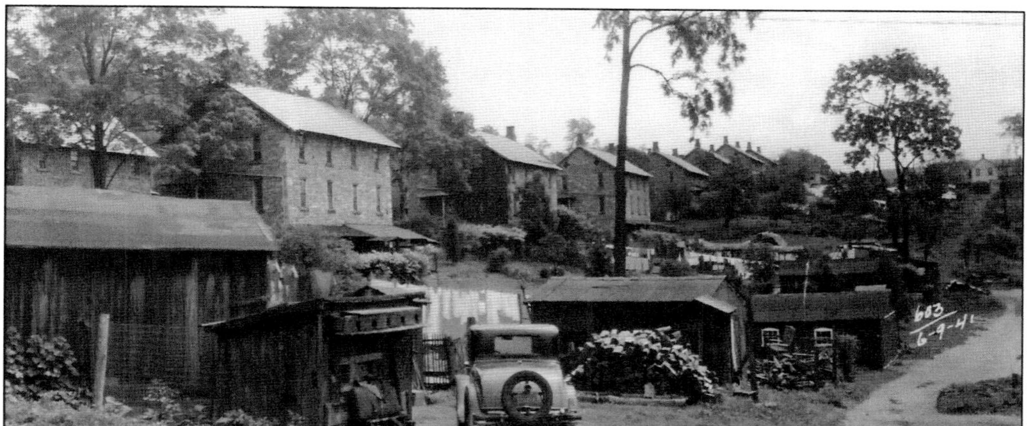

This 1941 photograph shows the backyards of Minersvillage. The gardens and woodpiles visible in the backyards provide a good idea of how people living in Cornwall got through hard times. (Courtesy of Daniel Chernich.)

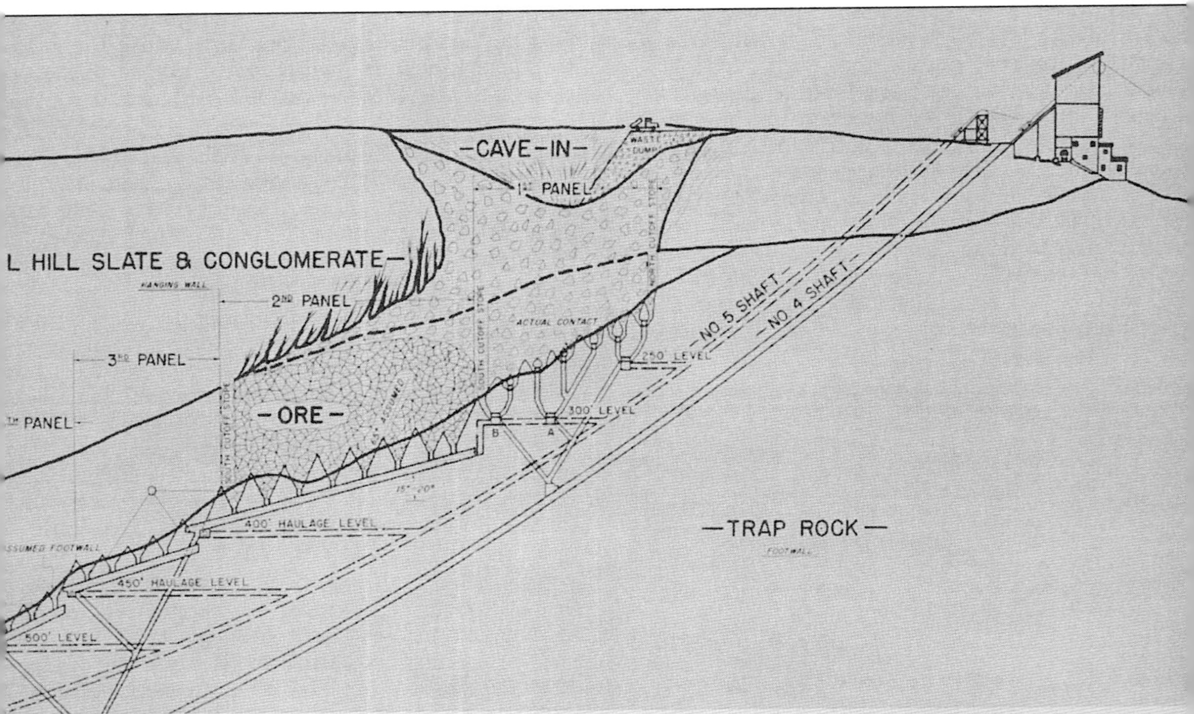

No. 4 mine is serviced by two inclined shafts, Nos. 4 and 5, sunk 200 ft in the trap rock footwall about 100 ft below the ore-trap contact. The slope at the collar is 36°, flattening to 26° at the bottom. Each shaft is 20 ft wide and 7 ft high with three compartments, two skipways, and a manway.

The No. 4 Mine was dewatered in 1936, and mining resumed in 1937. This diagram shows the plan for working No. 4 Mine. In 1940, No. 3 Mine was closed to prevent interference with the work in the open pit. This diagram first appeared in the July 1957 issue of *Mining Engineering* in an article written by R.G. Peets.

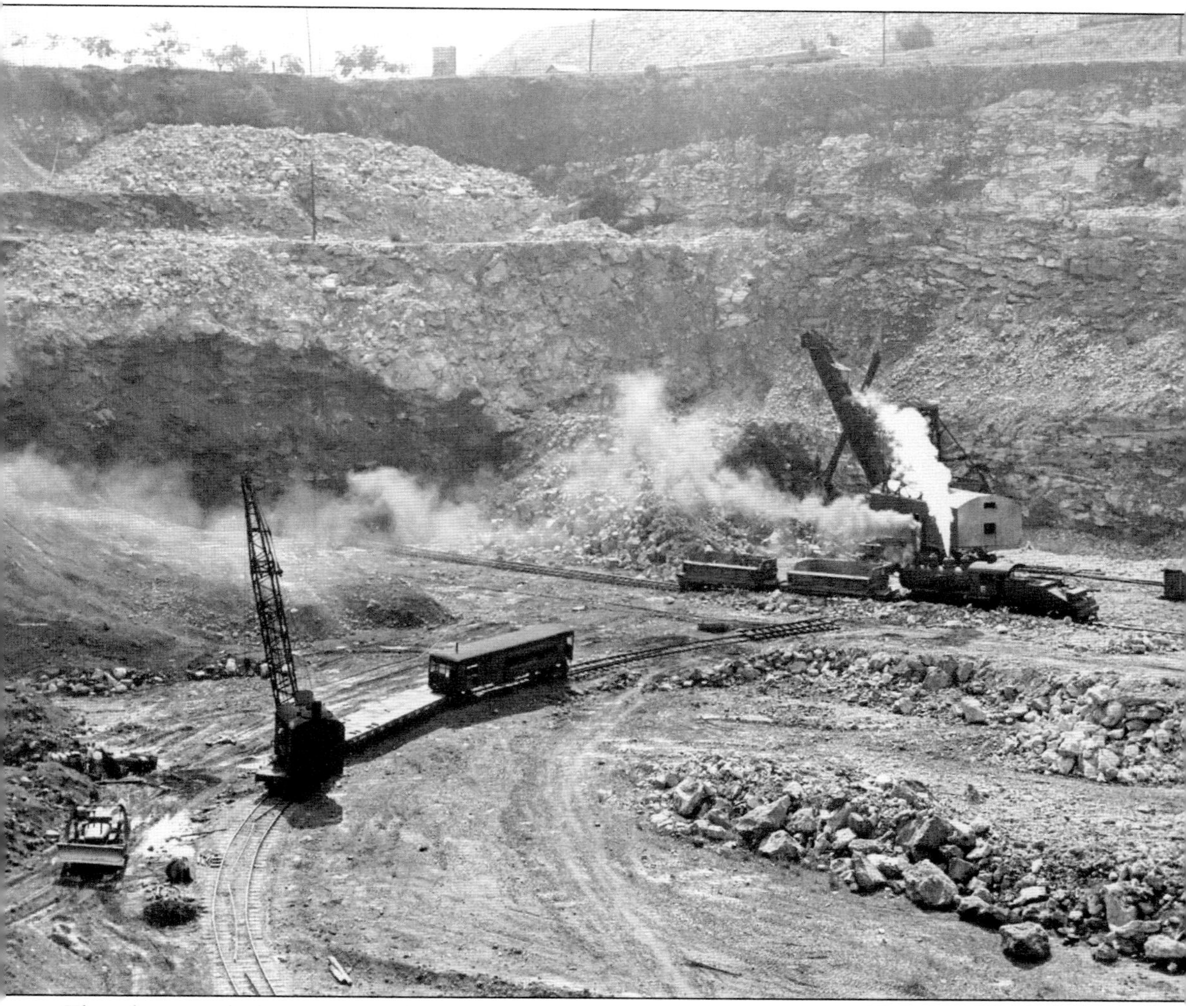

This photograph shows the continued mining operations in the open pit. The big electric shovel (right) loads the railcars, and a steam train can be seen. The lone passenger car in the center of the photograph was used to ferry miners to and from the mine. A steam engine would pull the passenger car loaded with miners into and out of the pit.

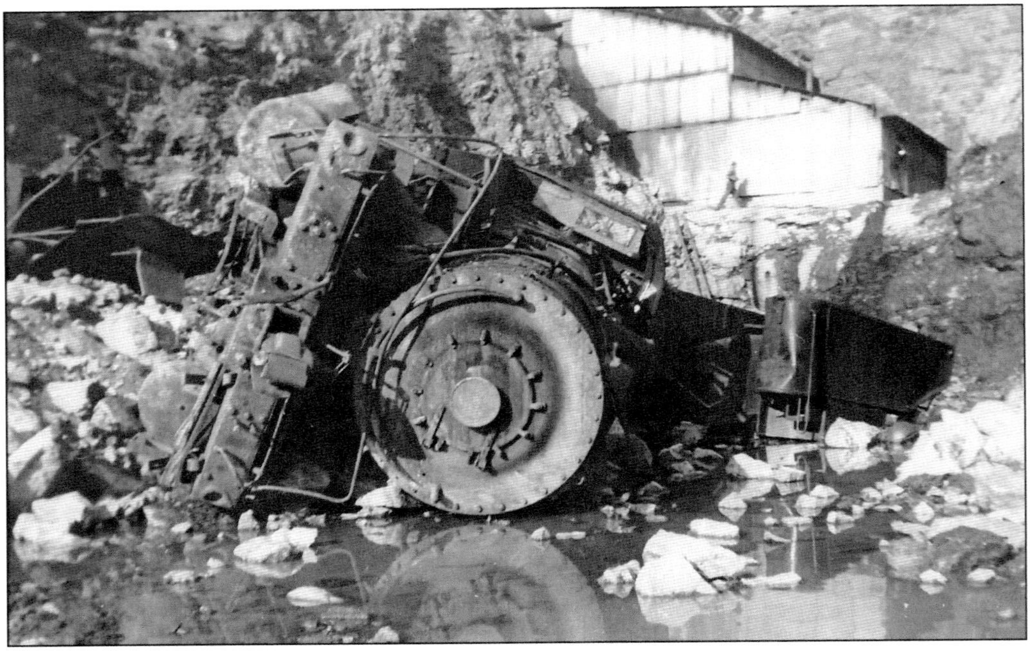

In 1941, a railroad accident occurred during the night shift. The train crew used lantern signals to communicate when an engine was coming down into the pit. Because of heavy fog, the signal was missed, and two engines—one exiting the pit and the other coming down the slope—collided. Both engines tumbled about 65 feet to the bottom of the pit, killing one engineer. The collision marked the last time lantern signals were used to communicate between trains at the mine.

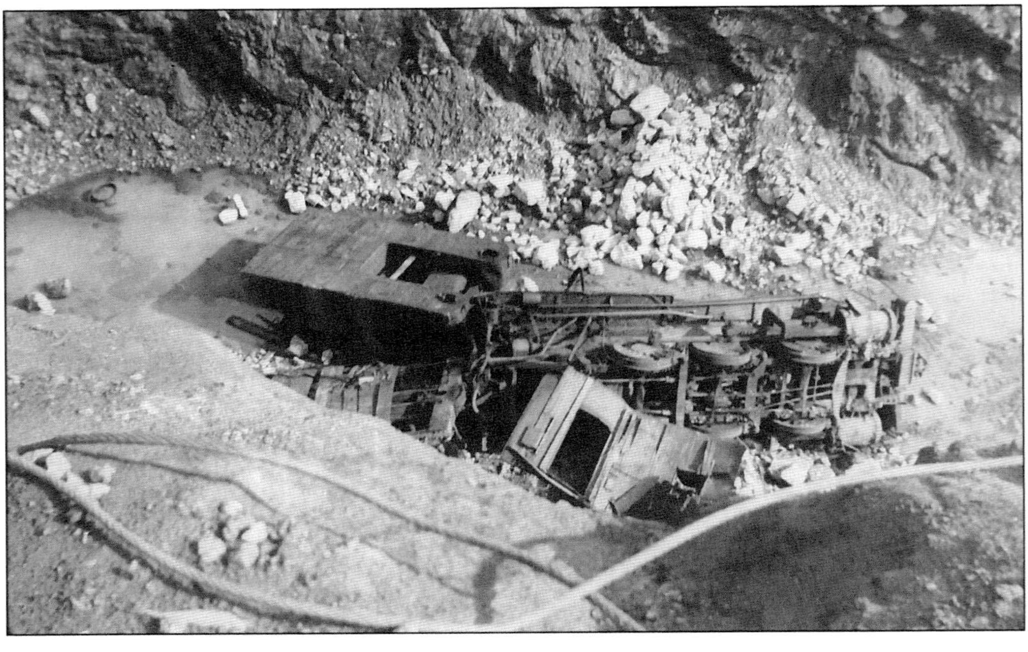

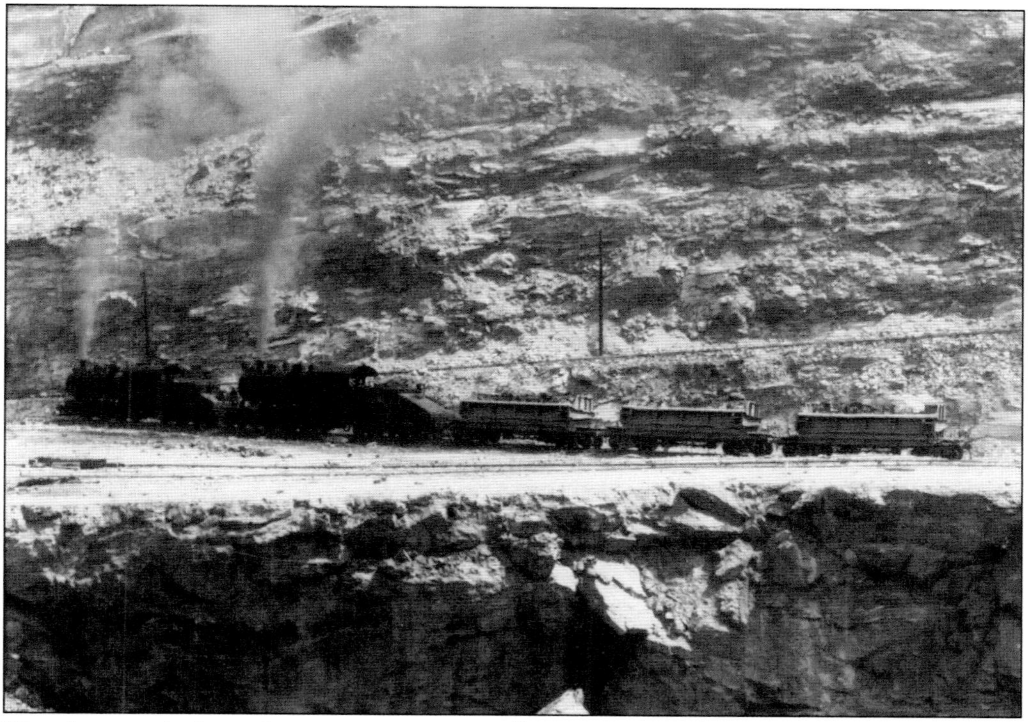

The photograph above shows steam engines pulling ore cars in the open pit. The image below shows steam engines moving railcars of ore in winter.

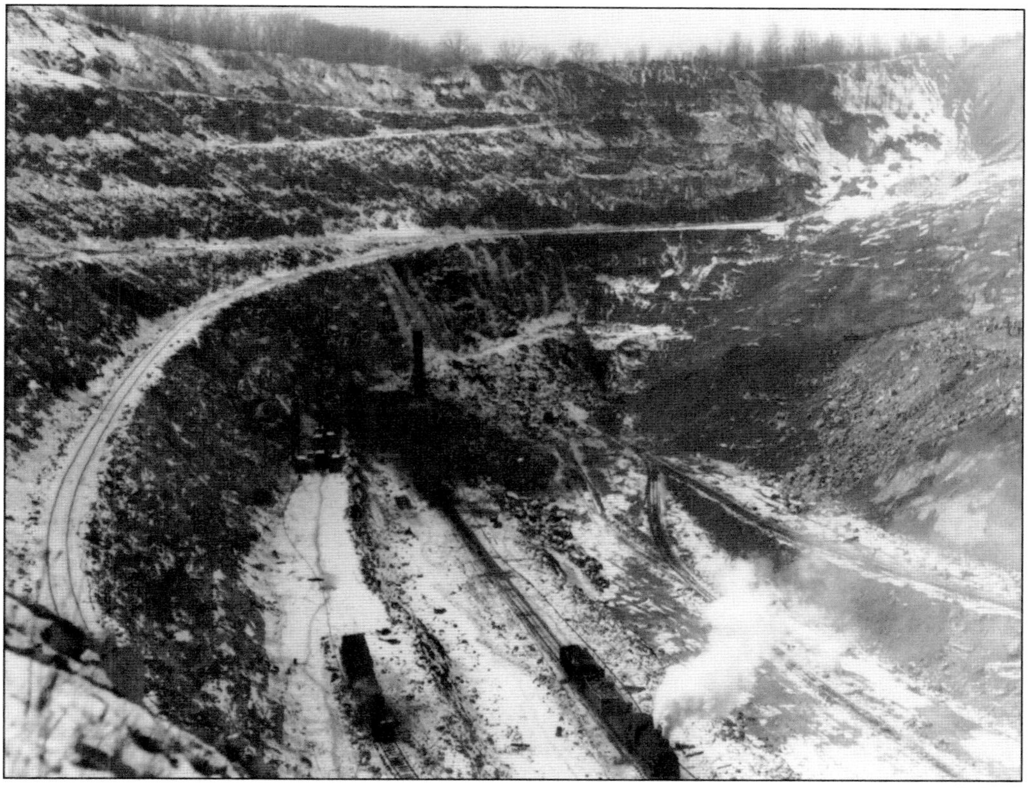

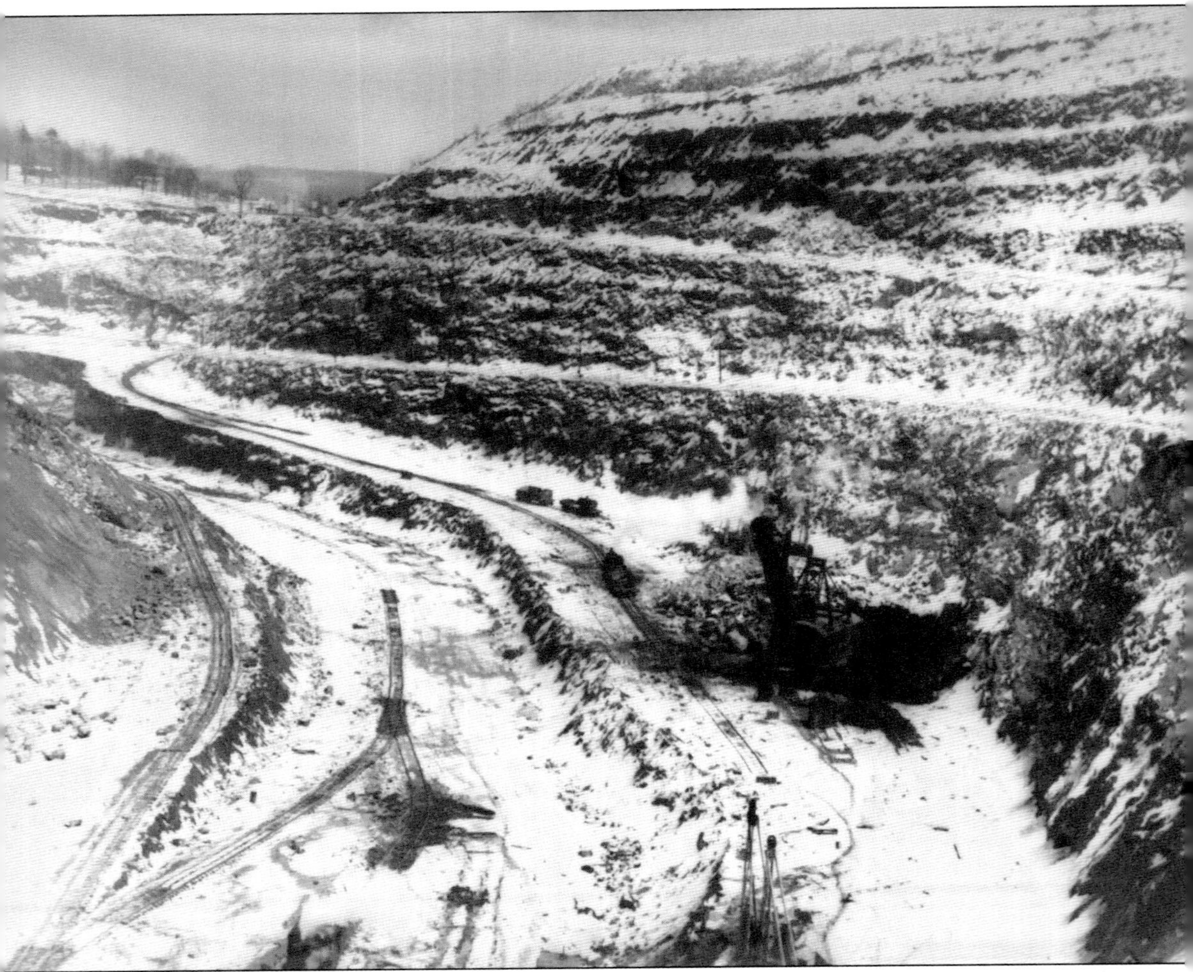

This photograph shows mining in the open pit during winter. Mining went on even in the harshest weather conditions.

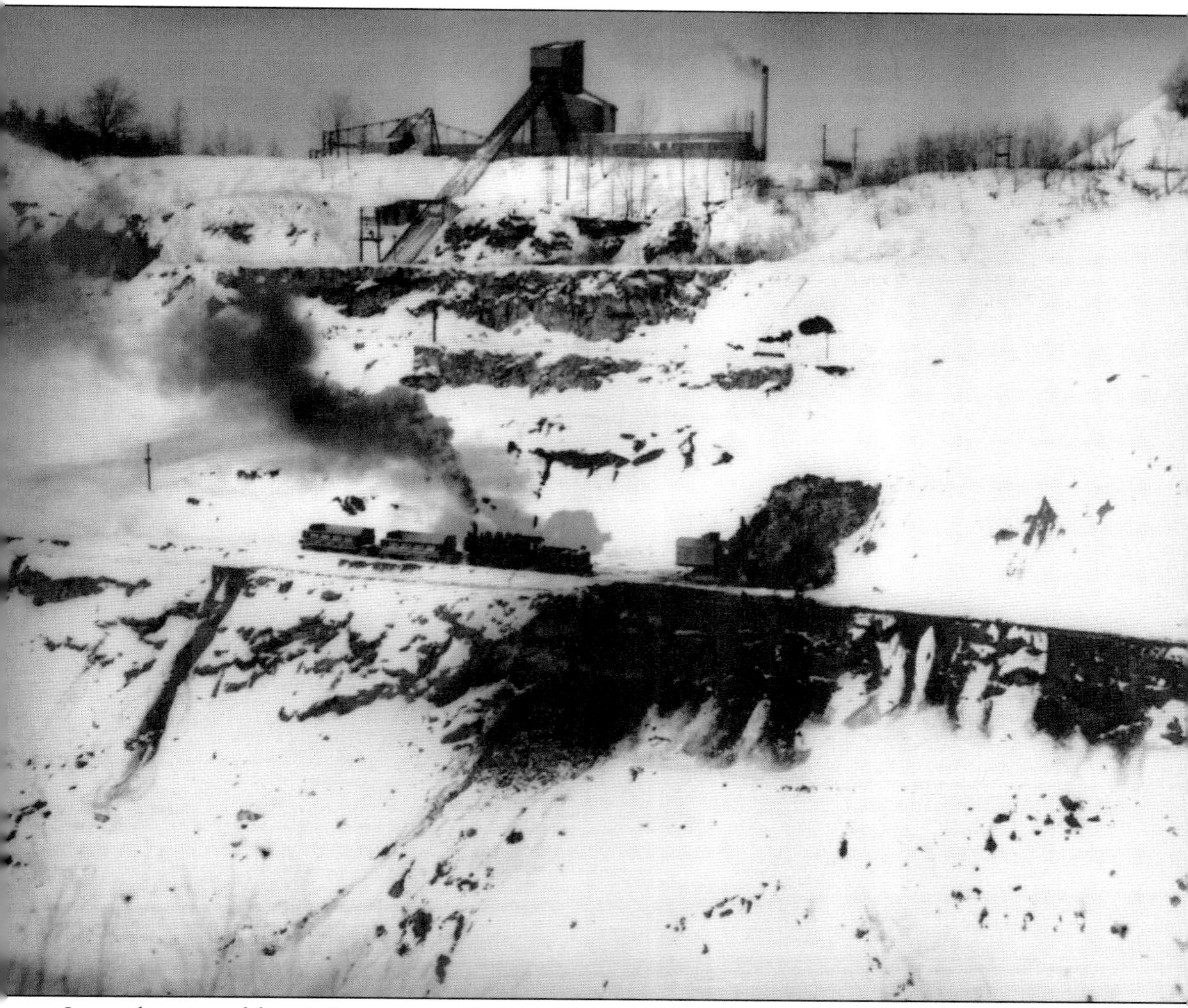

In another view of the mining operation during winter, a steam engine is pushing railcars away from the shovel that loaded them. The No. 1 Incline and headframe can be seen at the rim of the pit.

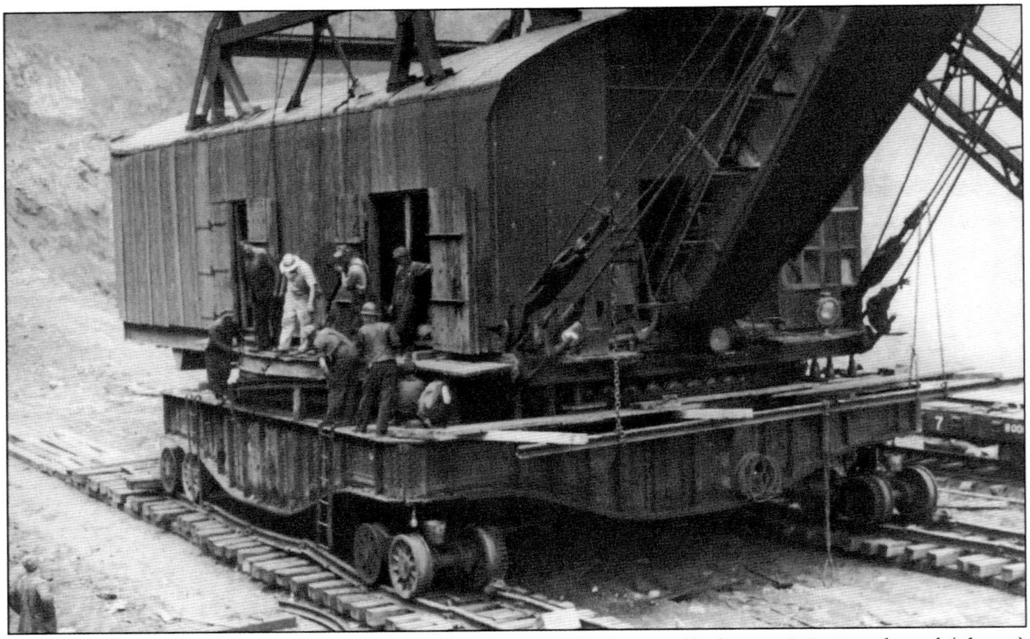

In 1944, big changes occurred in the open-pit mine. The large, all-electric Marion shovel (above) had worked itself to the bottom of the mine, where it was dismantled and sold. The old steam shovels in use at the time were replaced with two electric shovels. The smaller one had a dipper capacity of 1.25 cubic yards, and the dipper of the larger new shovel could hold three cubic yards. The steam engines of the Cornwall Ore Bank Railroad were replaced by six Mack dump trucks. The new vehicles were 200-horsepower, chain-driven, side-dumping workhorses. Below, one of these trucks is being dumped using the same equipment shown dumping railcars on page 39. The railroad tracks were also removed from the pit at this time.

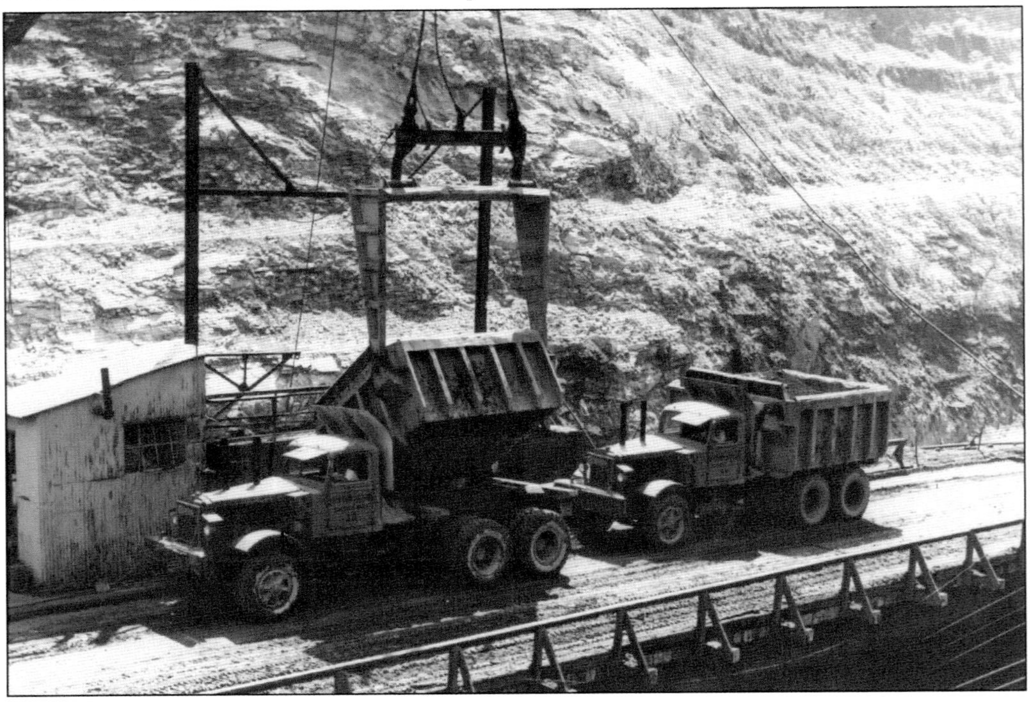

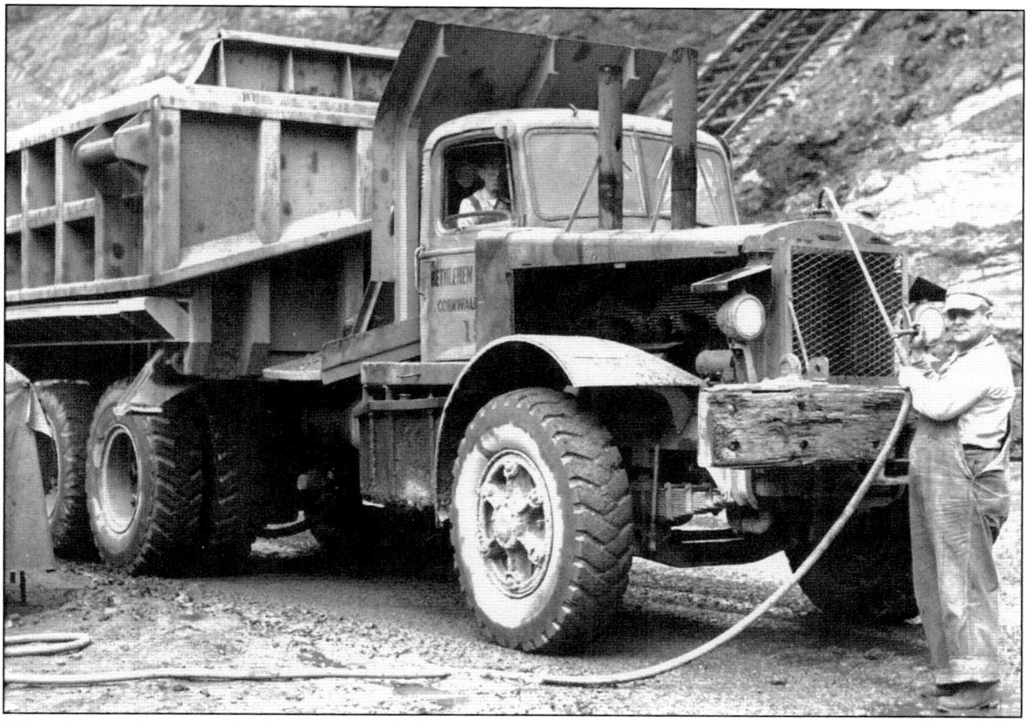

The above photograph shows one of the Mack dump trucks getting serviced. Below, a truck is being loaded by one of the smaller electric shovels. The steel at the dumping site appears at the bottom of the No. 1 Incline near the top of the picture.

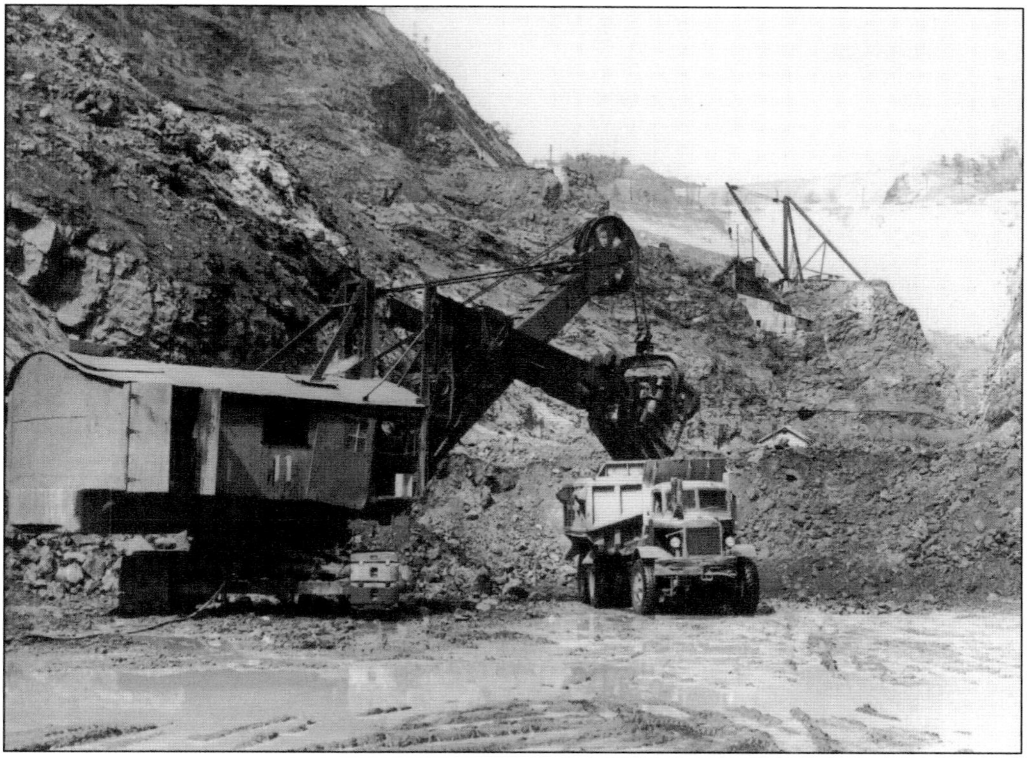

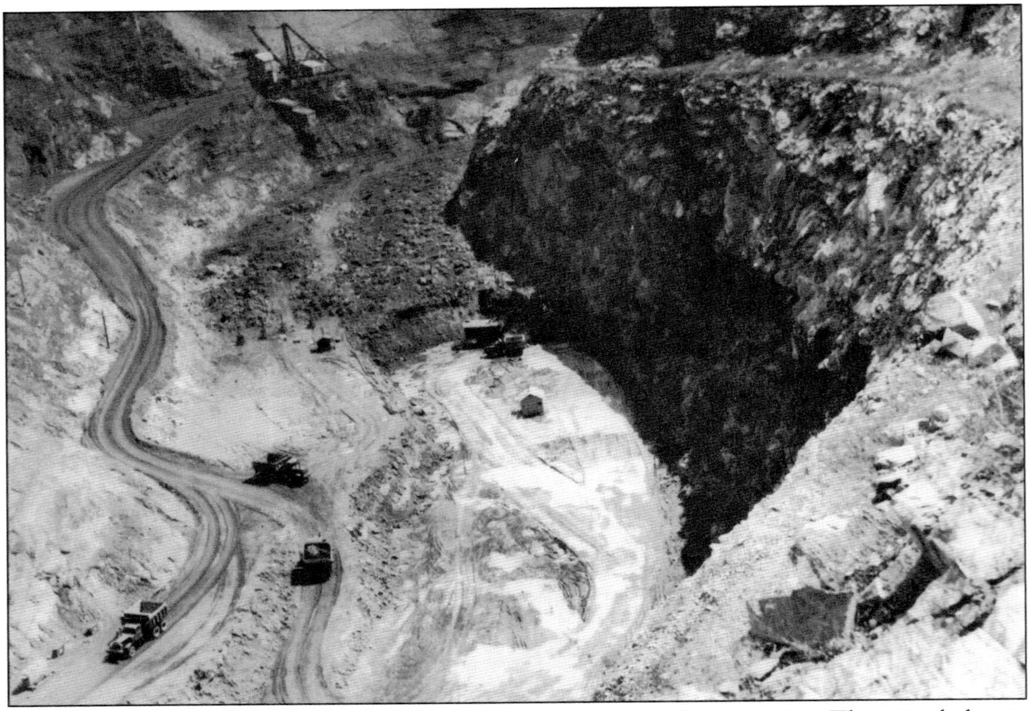

Above is the route these trucks took from the shovels to the dumping site. The view below is from the bottom of the pit. It shows how narrow the bottom became relative to the top. For safety purposes, the sides had to be angled so that the bottom became narrower as the depth increased. This picture was taken in the late 1940s when the open-pit operation was coming to a close. The No. 3 Mine was active at this time to mine the ore around the pit that could not be extracted using the open-pit method.

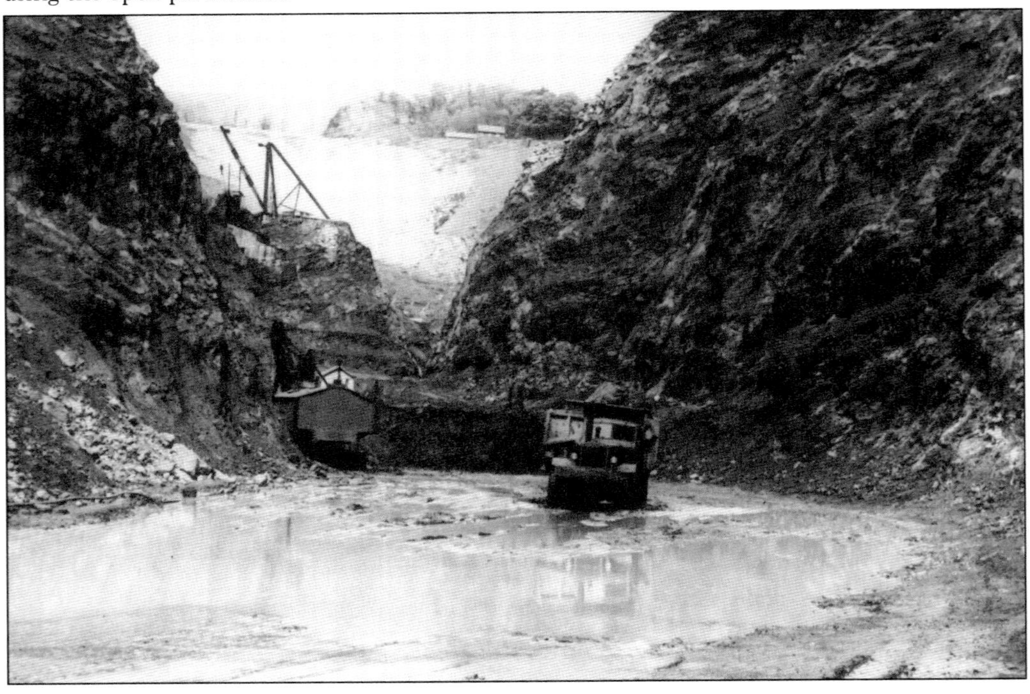

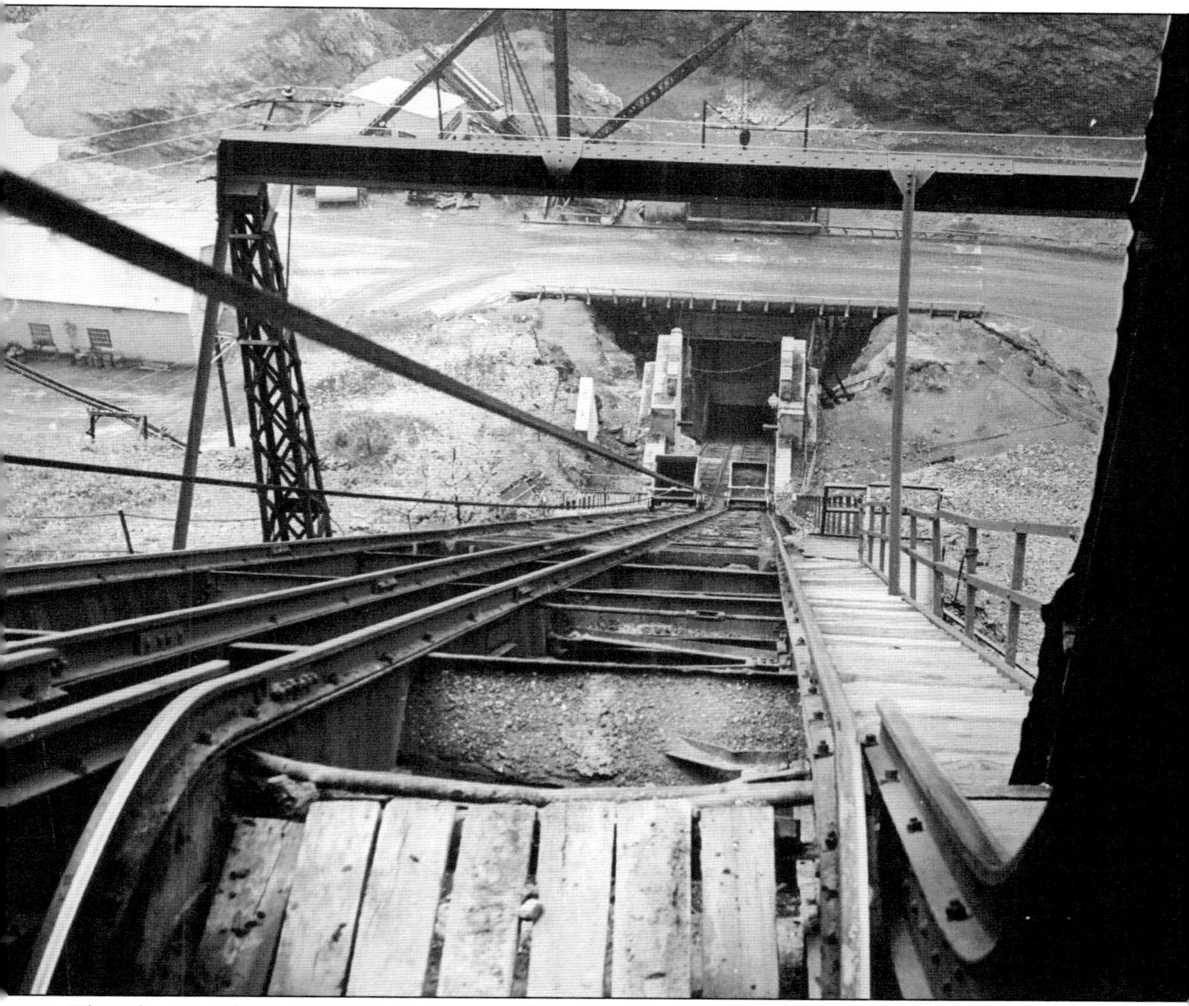

This photograph was taken from the top of the No. 1 Incline. The top of the picture shows the depth of the open pit. In the upper center are the roadway the trucks traveled on the way to dump their loads into the crusher and the crushed material being brought to the surface in the skips for processing.

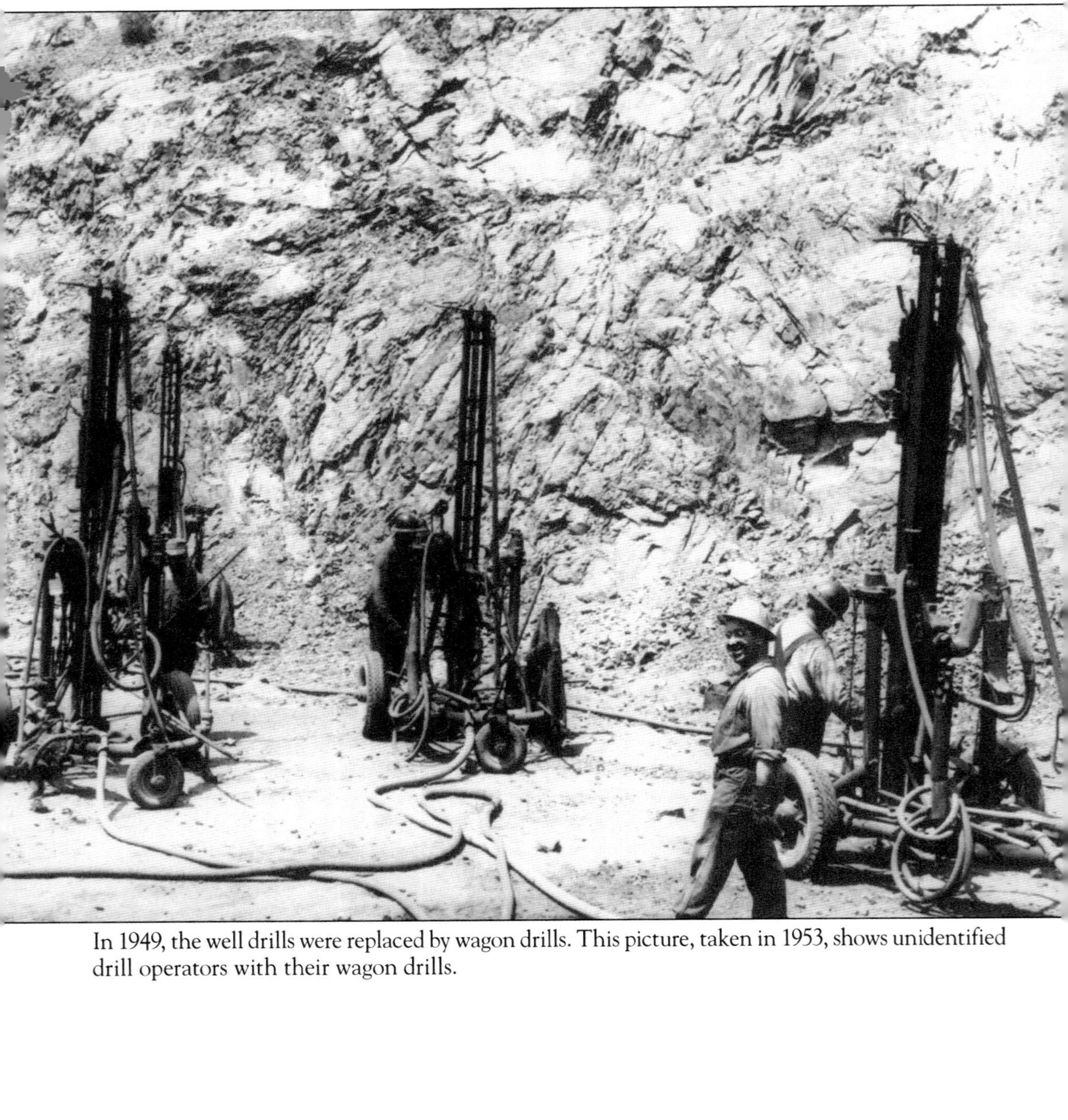

In 1949, the well drills were replaced by wagon drills. This picture, taken in 1953, shows unidentified drill operators with their wagon drills.

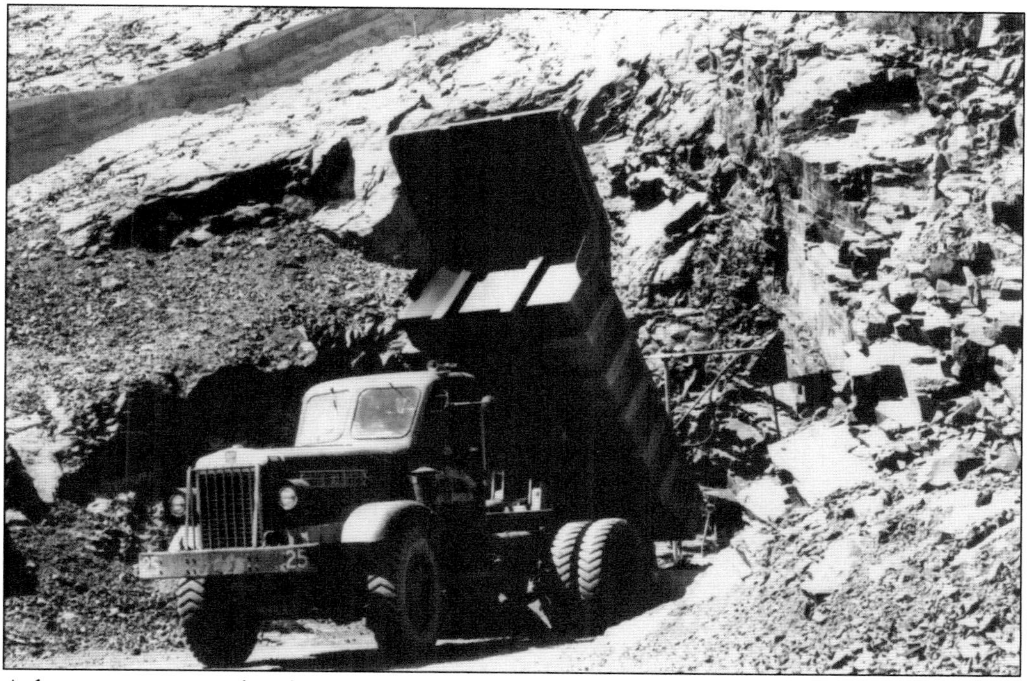

A dumpsite was created in the open pit. This truck is actually dumping into the No. 3 Mine, which is 12 feet below the bottom of the pit. In the underground mine, the ore would be transferred to the skips that transported the material out of the mine. In 1953, however, after 216 years of open-pit mining, work in the pit finished. Any and all equipment was removed and sold. The photograph below shows how empty the once-busy pit was after mining operations there ceased.

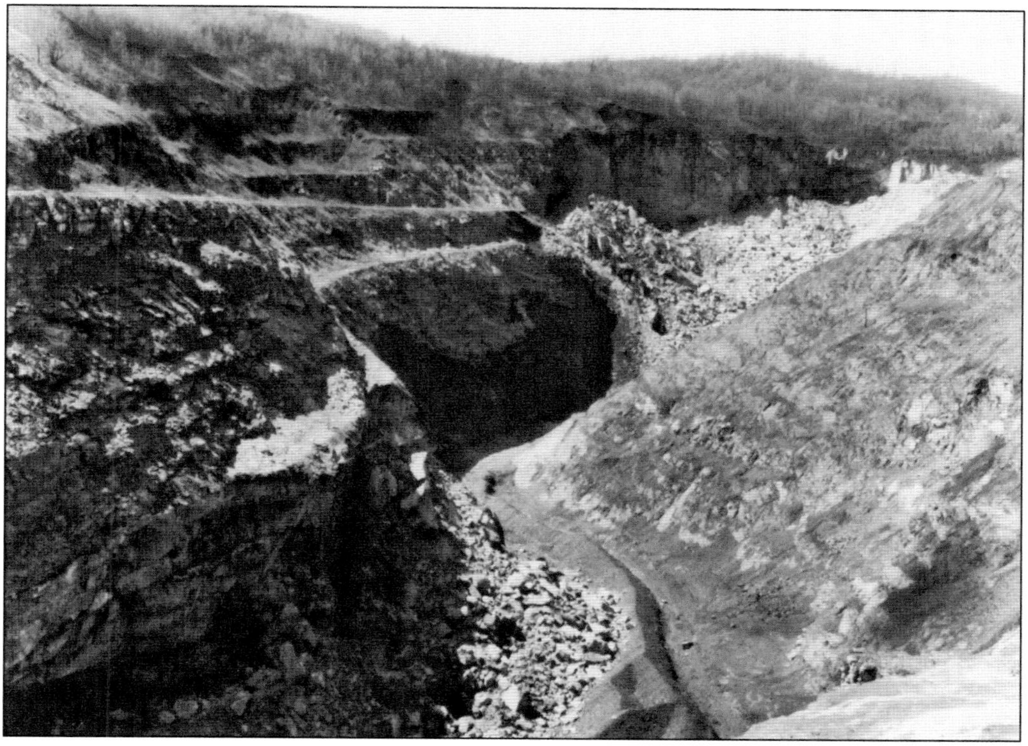

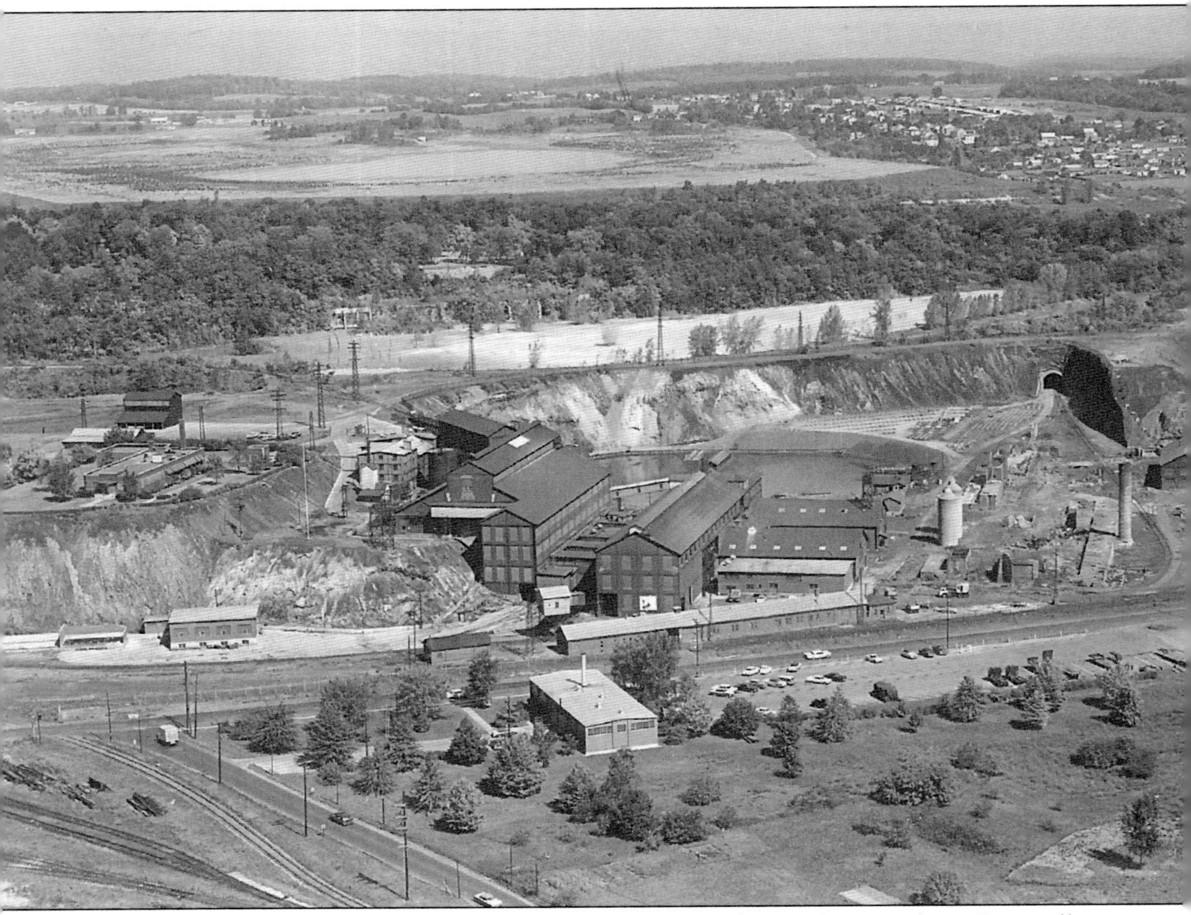

The Lebanon Concentrator Plant was in Lebanon, Pennsylvania. Iron ore from Cornwall was shipped to this plant for processing. The plant used a flotation process, which had been adapted in 1912. As the name implies, raw material was separated using liquid. Pyrite and chalcopyrite floated away from the heavier iron. The pyrite-chalcopyrite phase contained gold, silver, and copper. General superintendent of the mines Milan Lipensky reported that 11,000 pounds of copper, 5,000 ounces of silver, and 100 ounces of gold were separated from the pyrite-chalcopyrite phase in a typical month. In about 1940, cobalt was also discovered in this lighter phase. A *Lebanon Daily News* article from 1953 details mining in Cornwall: "Reports of the US Bureau of Mines reveal that Cornwall, for years the only producer of cobalt in the United States, is still leading the list despite the opening of other operations in the West. For some years the production has averaged 600,000 pounds per year, Bureau reports on Pennsylvania show—and Cornwall is the only mine in the state where cobalt, as well as silver, gold, and copper are byproducts. So Pennsylvania statistics for these metals are Cornwall statistics."

Five

Churches, Businesses, and the Community of Cornwall

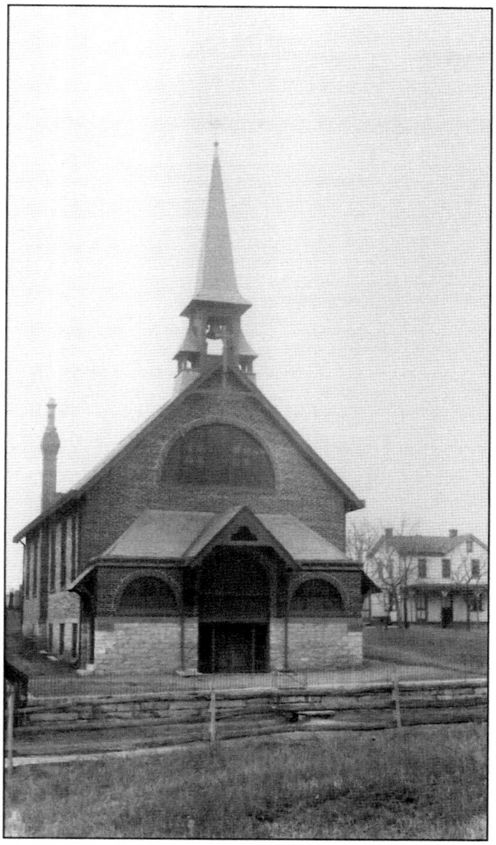

The Sacred Heart of Jesus Catholic Church was built in 1889 at this location along the Cornwall Pike (formerly the old plank toll road). In 1914, the church building was dismantled and moved nine-tenths of a mile to its current location in Cornwall Center. A picture of the church in the new location is on page 84. (Courtesy of Daniel Chernich.)

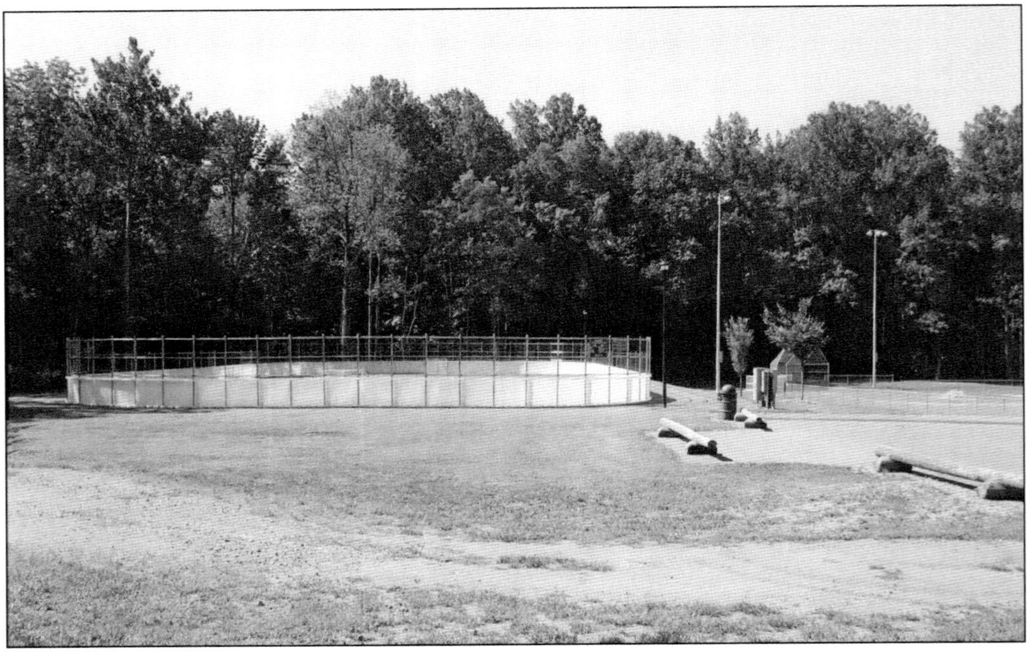

Above is a current picture of what is to be Ferndale Park. It is located east of Minersvillage along the old US Route 322. Ferndale Park now is home to a Little League baseball field and a street-hockey rink. The Minersvillage Mennonite Church was established there in 1931. The church held services in the Minersvillage dance hall. In 1936, the dance hall burned to the ground in a spectacular fire. The church purchased the Minersvillage Boarding House (below) in 1937. The building was used exclusively as a church until 2003. (Below, courtesy of Daniel Chernich.)

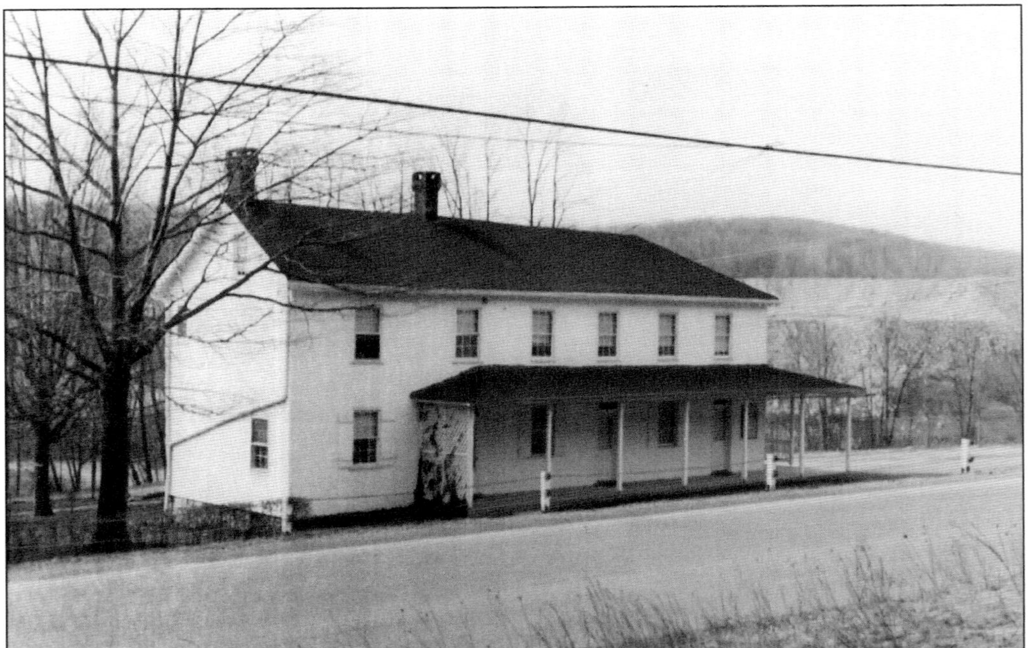

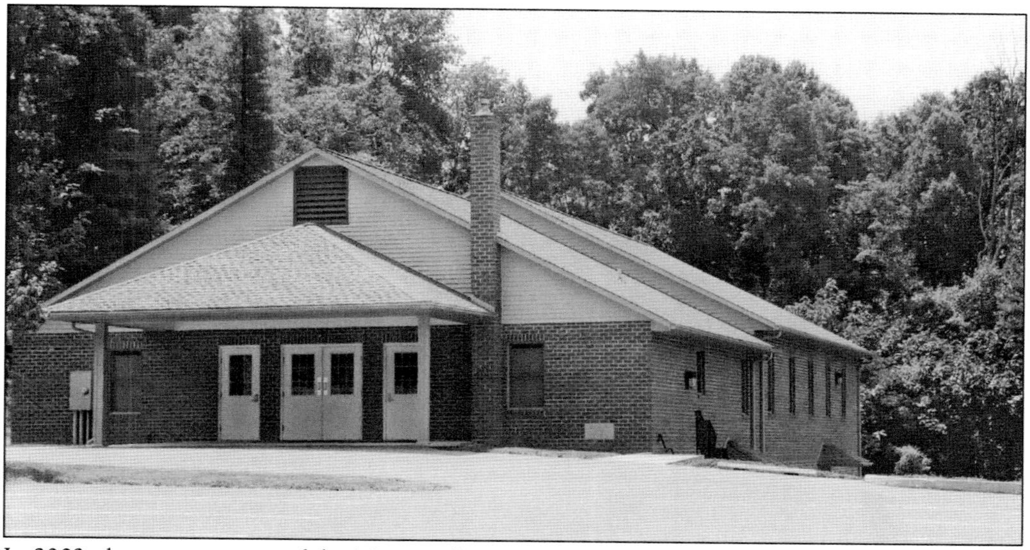

In 2003, the congregation of the Minersvillage Mennonite Church built a new house of worship west of Minersvillage and the open-pit mine.

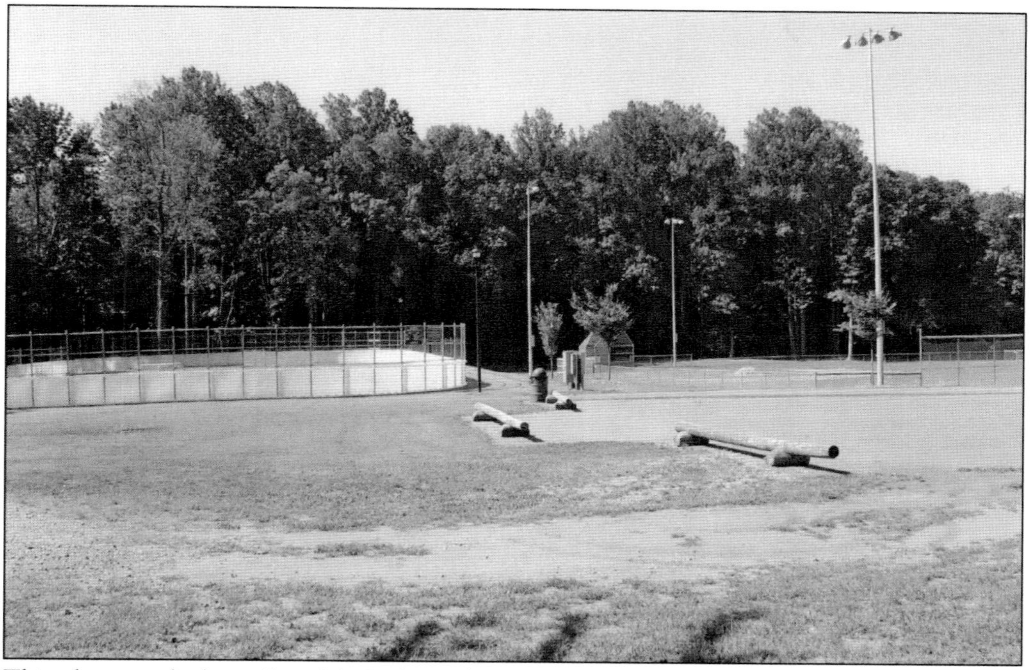

This photograph shows Ferndale Park on the eastern end of Minersvillage as it appears today. The highway going straight through Minersvillage was US Route 322, and the park was located beside the road. During World War II, troops would use Route 322 to travel to the coast to change duty stations or for deployment overseas. Occasionally, convoys would stop to eat and set up camp for the night in the park. If the weather were harsh and the group small enough, some of the families in the village would invite soldiers into their homes so they could keep warm during the night. Many of these families in the village had loved ones serving in the military during the war, and this was a way for them to support the soldiers. During the 1950s and early 1960s, Boy Scout troops would camp at the park. The Scouts would go rock hunting on the dumpsites to earn merit badges for geology.

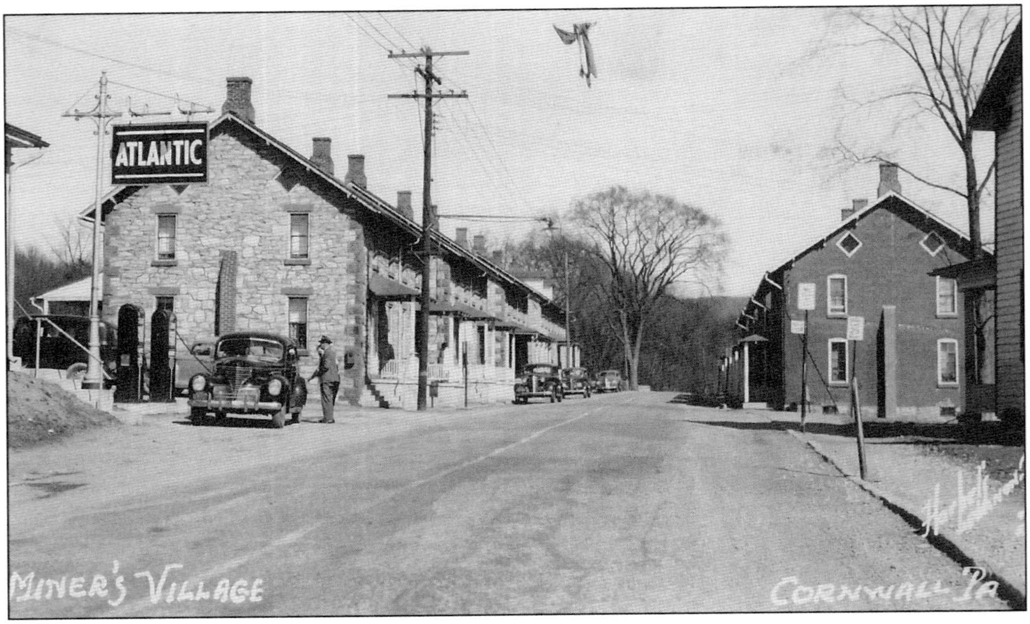

Both pictures show the Minersvillage Store and vicinity. The 1940s postcard above is of the gas pumps out front and the east-west Route 322 (now Boyd Street) through the village. In the 1950s image below, taken from across Route 322, is the store itself, where meats, cheeses, canned goods, in-season vegetables, penny candy, and ice cream could be purchased. There was also a small soda fountain in the rear of the business. The road to the right is Shirk Street. Today, the store building is a private residence. (Below, courtesy of Daniel Chernich.)

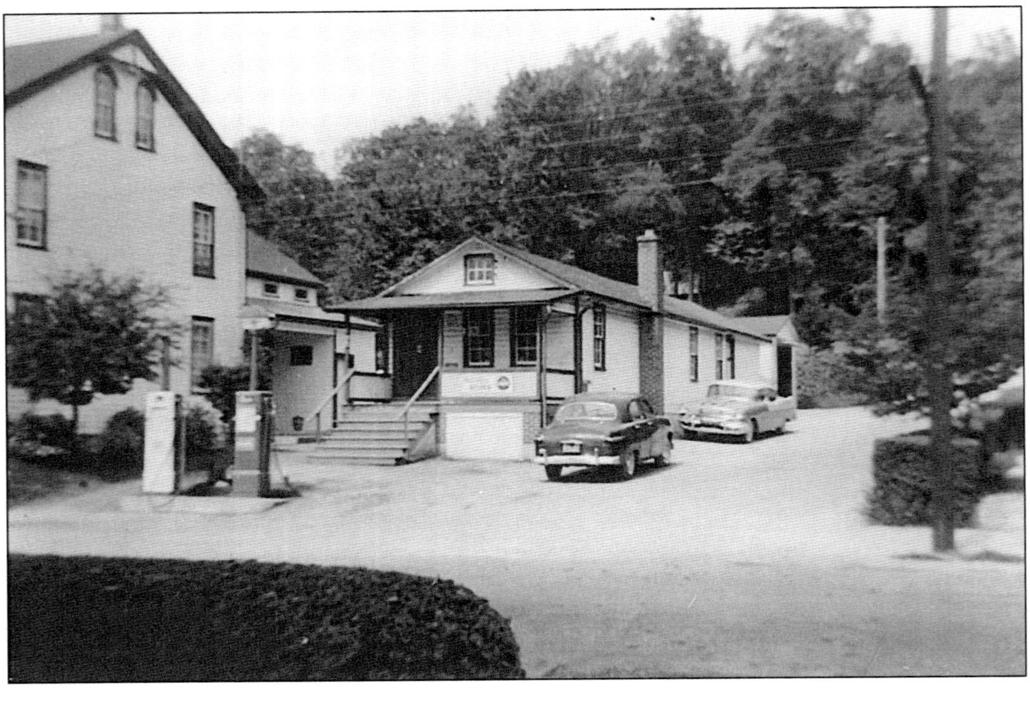

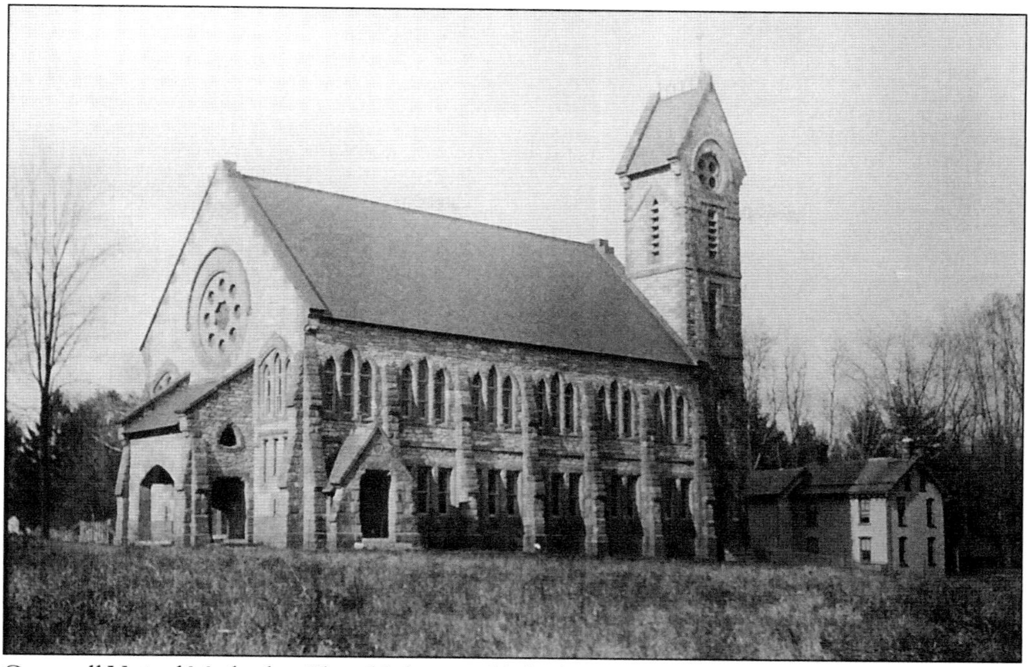

Cornwall United Methodist Church's first established congregation on record was in 1793. Its first Sunday school was held in the old mill house (later known as the Cornwall Store and pictured on page 84). In 1834, the congregation constructed its first church building of brick. Over the next 40 years, the congregation outgrew that structure, and in 1878, the stone church shown in the image above was completed. The brick church was torn down. In 1998, the congregation had again outgrown its building, making an addition to the church necessary. The photograph below shows the church as it stands today. (Above, courtesy of Cornwall Manor Manypenny Library.)

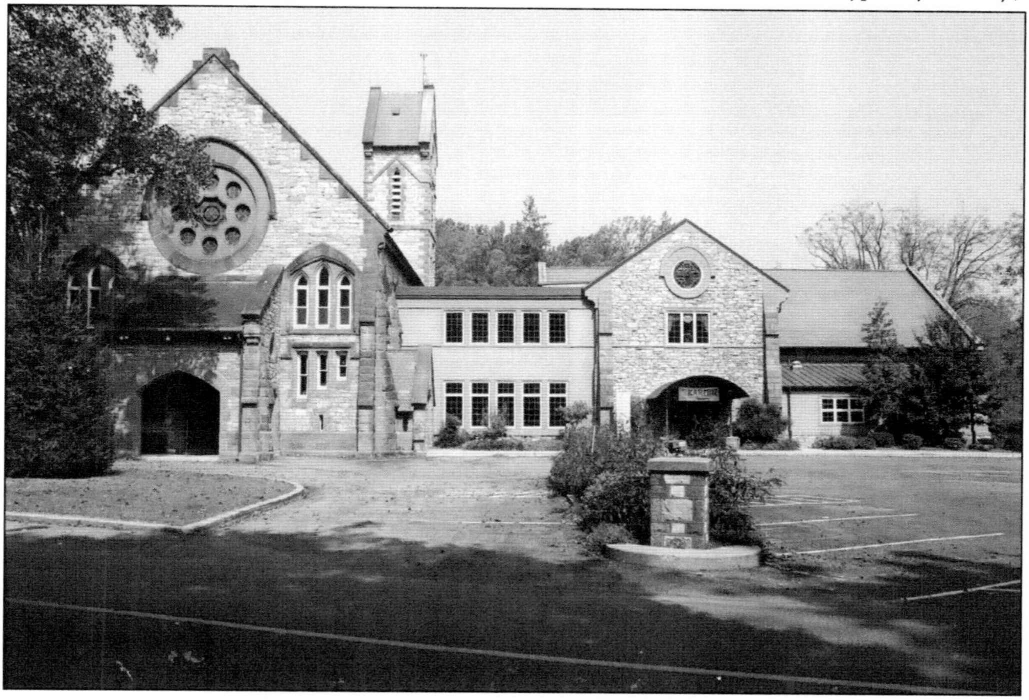

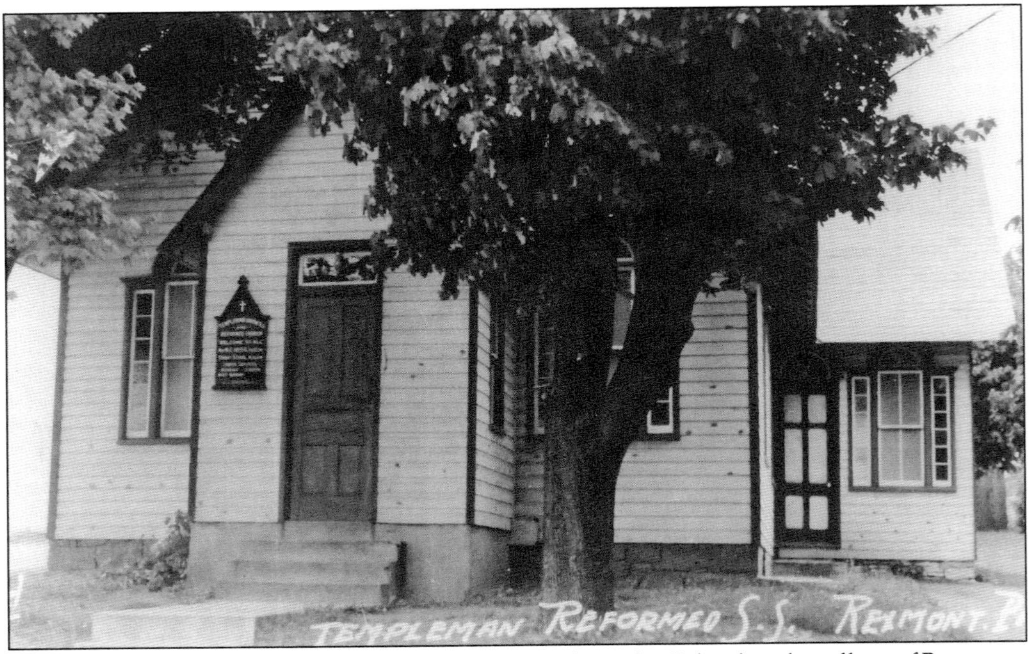

The building above housed the Templeman Reformed Sunday School in the village of Rexmont. In about 1949, this structure was moved to the east and set on a foundation with a basement, as shown in the picture below. The congregation chose to become part of the United Church of Christ in Quentin in 1971. This structure was sold in 1973 and became a private residence, which it remains, although a bed-and-breakfast now operates out of it as well. (Both, courtesy of Mr. and Mrs. Ray Dishong.)

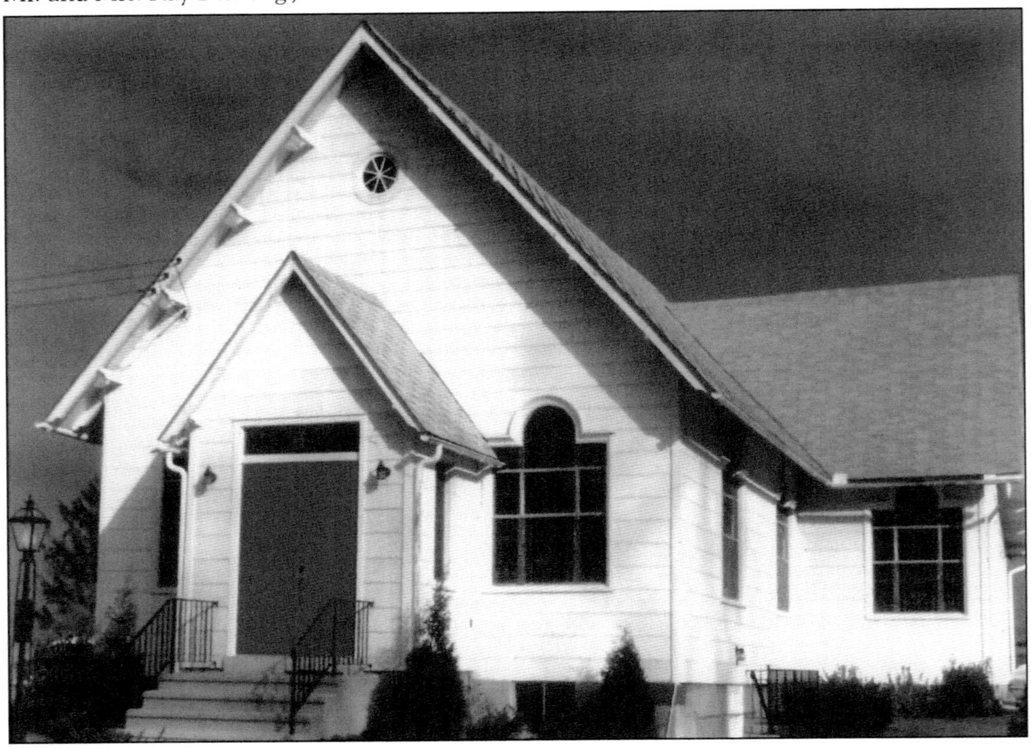

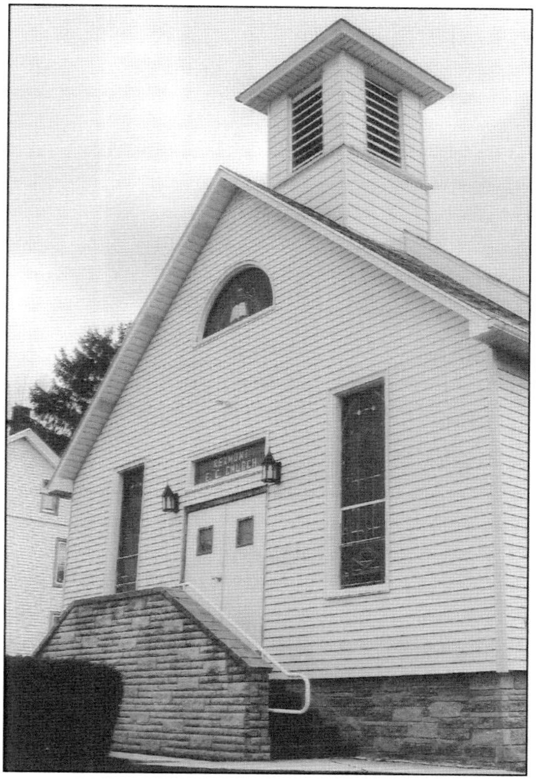

At right is the Evangelical Congregational Church in the village of Rexmont. This church was built in 1895 and has been remodeled repeatedly during the 115 years it has served its congregations. The image below shows one of its very beautiful stained glass windows.

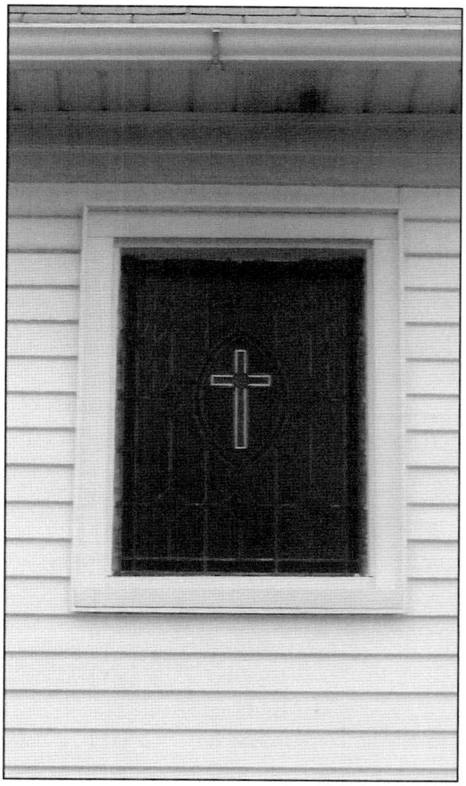

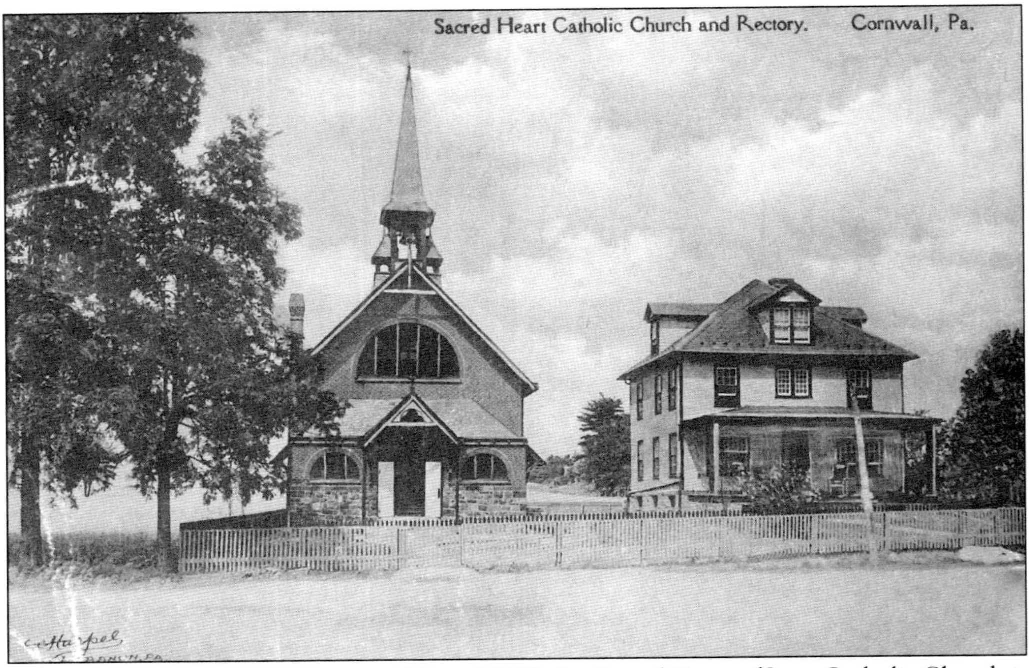

Seen in this postcard from between 1914 and 1920, the Sacred Heart of Jesus Catholic Church is situated at a prominent location in Cornwall Center. The church was dismantled at its original spot and moved here in 1914. Its beautiful stained glass windows were taken out and reinstalled. A rectory was built at that time as well. Today, the structures appear much the same.

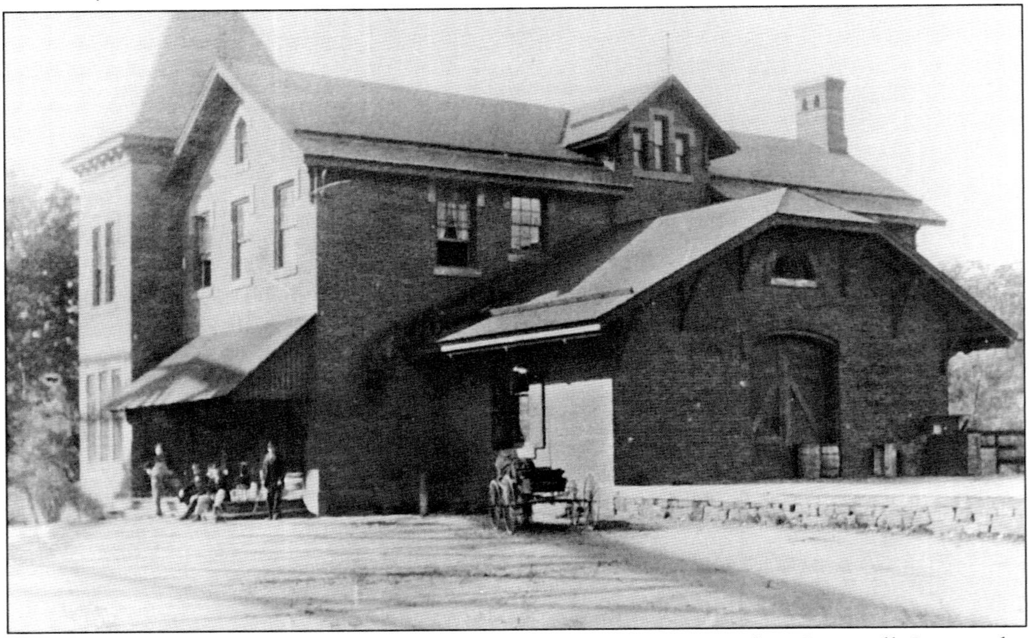

The Cornwall Store housed the community's first post office. Located in Cornwall Center, the building was originally used as the mill house for the mill shown on page 85. The first floor was the actual store where clothing, shoes, housewares, and meat could be purchased. The second floor was used as living quarters. In later years, the store would deliver orders to homes. This building now houses a bed-and-breakfast known as the Cornwall Inn.

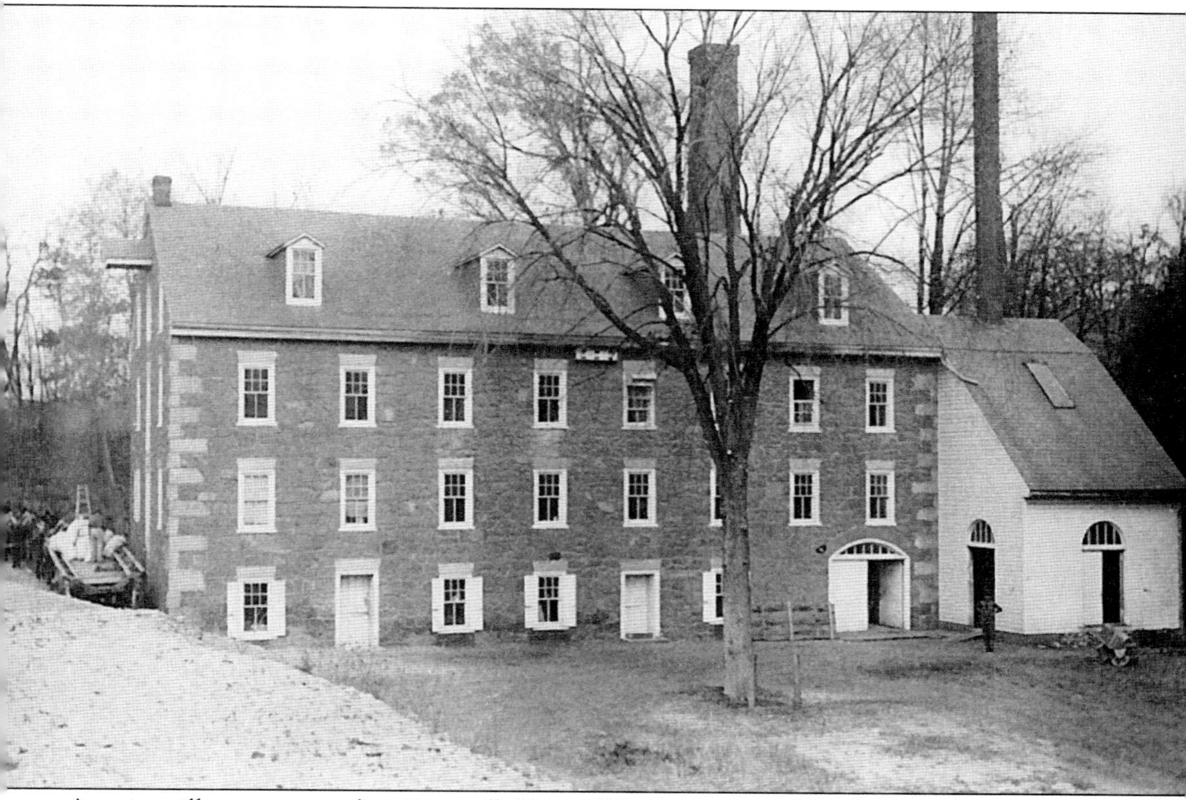

A grain mill once operated in Cornwall. The mill house is pictured on the preceding page. A short race was built to divert water from a creek to turn the waterwheel and power the mill. The waterwheel was housed in the frame section of the building. A team of mules and wagon used to haul away grain can also be seen. This building was razed in the early 1950s.

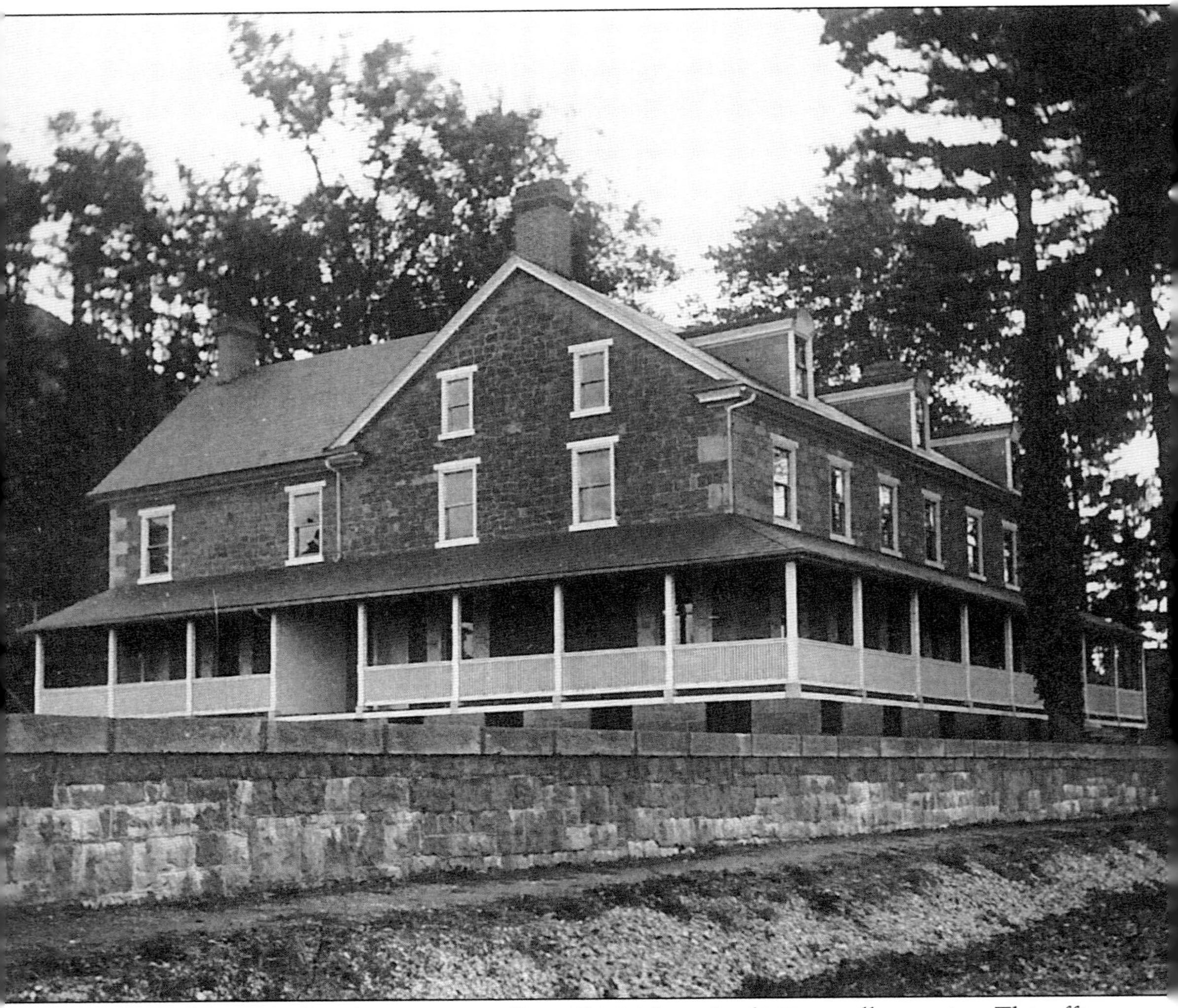

This building was the administrative office of Bethlehem Steel's Cornwall operation. This office managed the mines, the open pit when it was operating, the pellet plant, real estate rentals, and property upkeep. (Courtesy of Cornwall Manor Manypenny Library.)

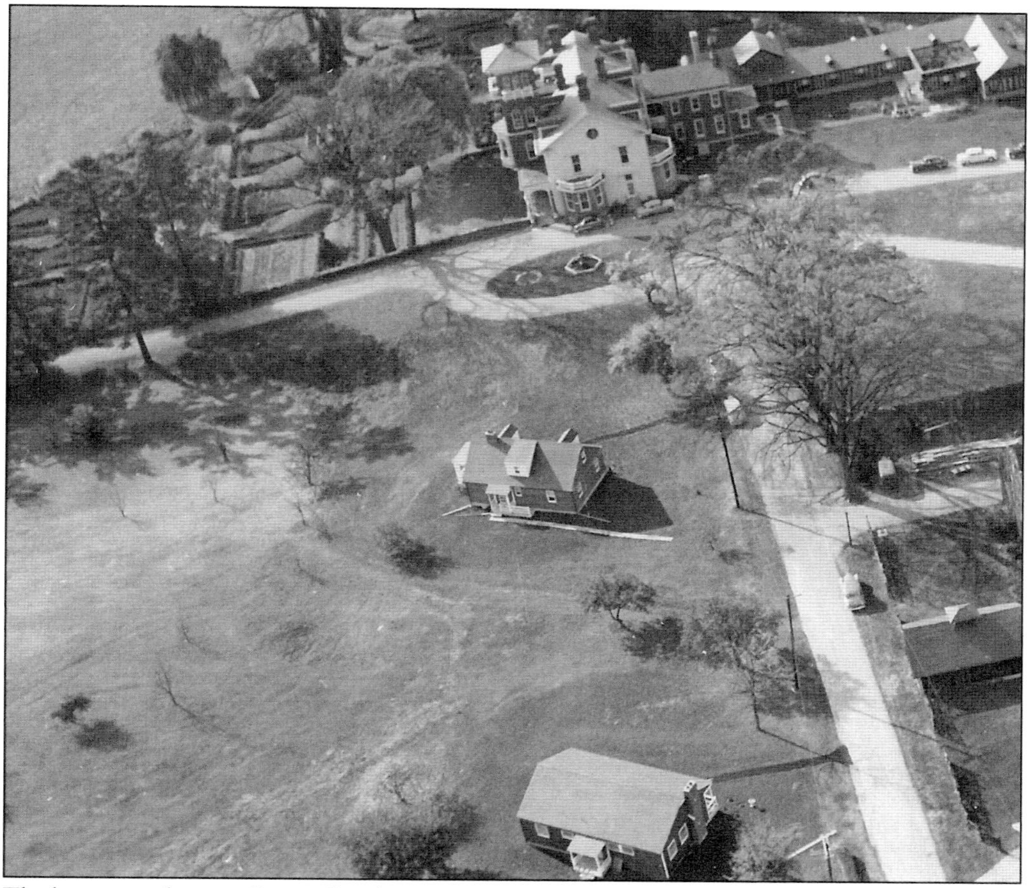

The largest employer in Cornwall today is Cornwall Manor, a continuing care retirement community started in February 1949. The first four residents arrived in October 1949. Located on an 85-acre campus that was once the iron master's estate, Cornwall Manor now employs 325 people, 260 of whom are full-time. Many of the old buildings have been repurposed for retirement living. The photograph above shows the iron master's mansion. An annex has been added to the rear of the home. To the left of the mansion are the formal gardens. In the picture below, the old stable is on the left. Today, residents occupy all of these buildings, which have been beautifully redesigned as apartments. (Both, courtesy of Cornwall Manor Manypenny Library.)

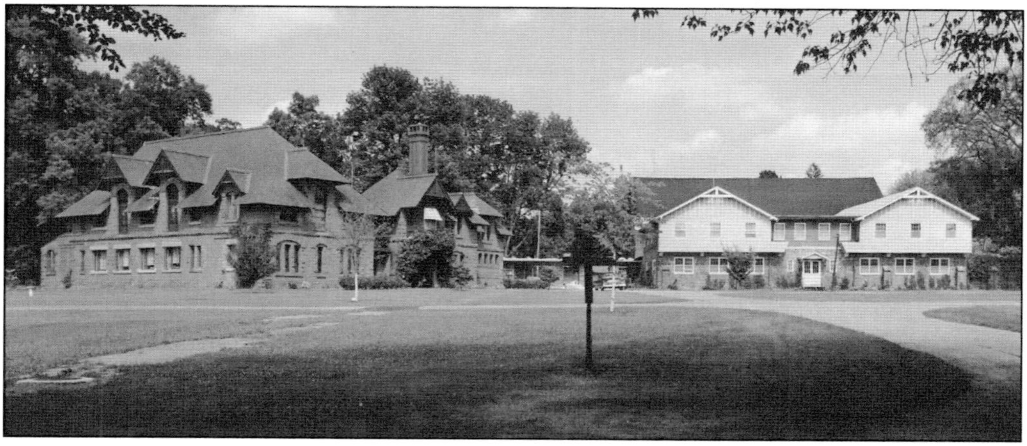

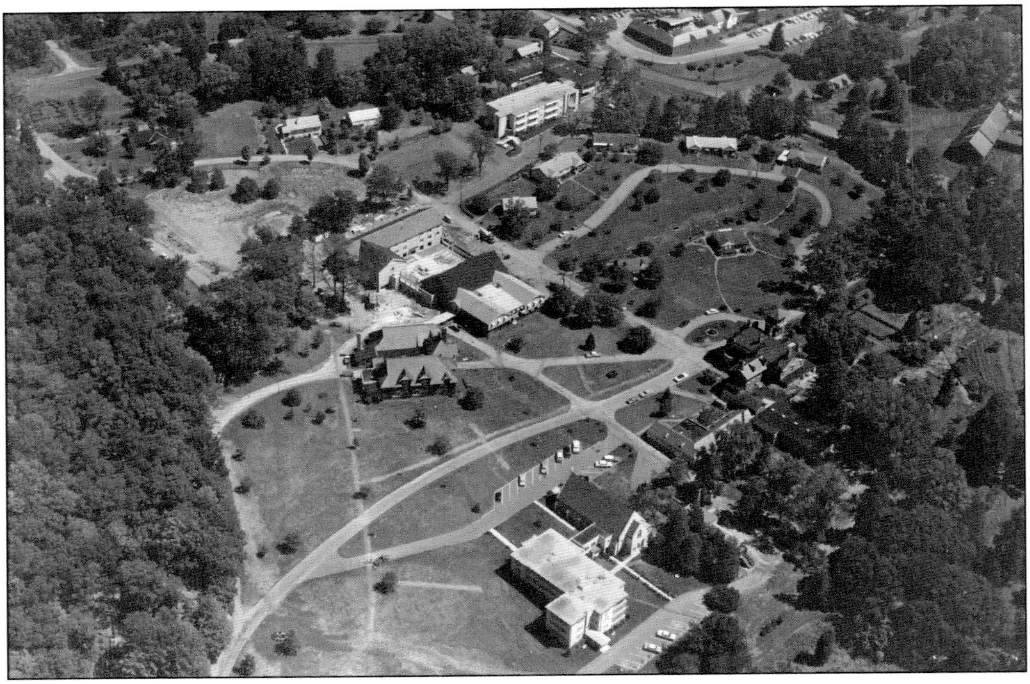

The Cornwall Manor campus is large and best viewed from the air. The picture above shows the variety of living quarters offered to residents. Among the apartments and houses sits the chapel, almost at the center of the photograph with a sharply sloped roof. The photograph below shows the gatehouse, also a resident's home, located on the edge of the campus. (Both, courtesy of Cornwall Manor Manypenny Library.)

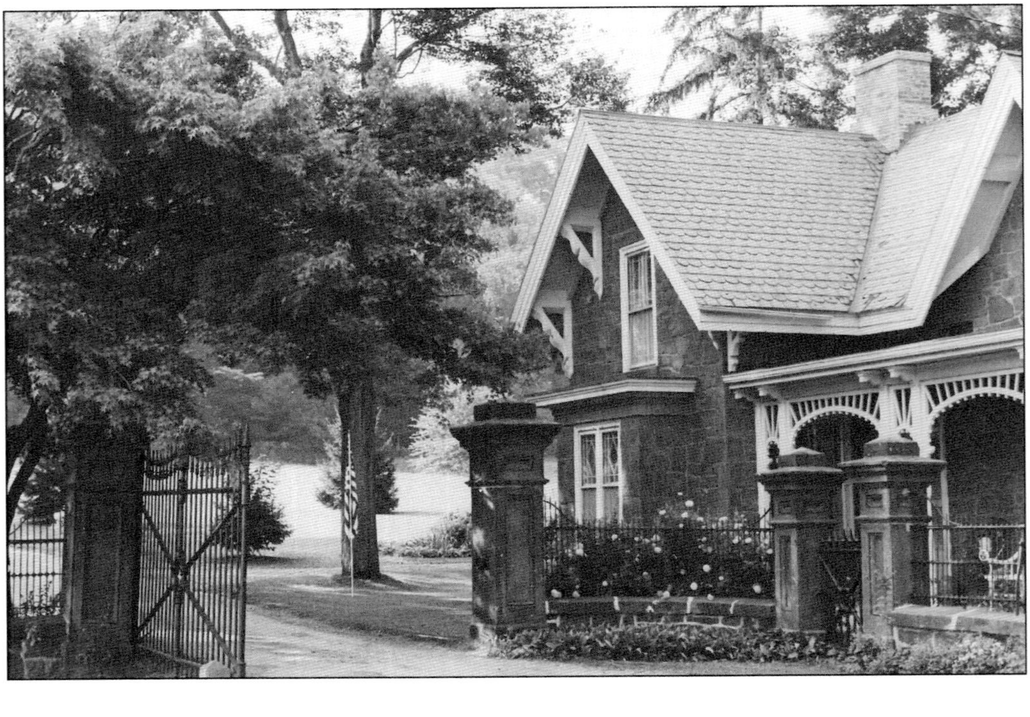

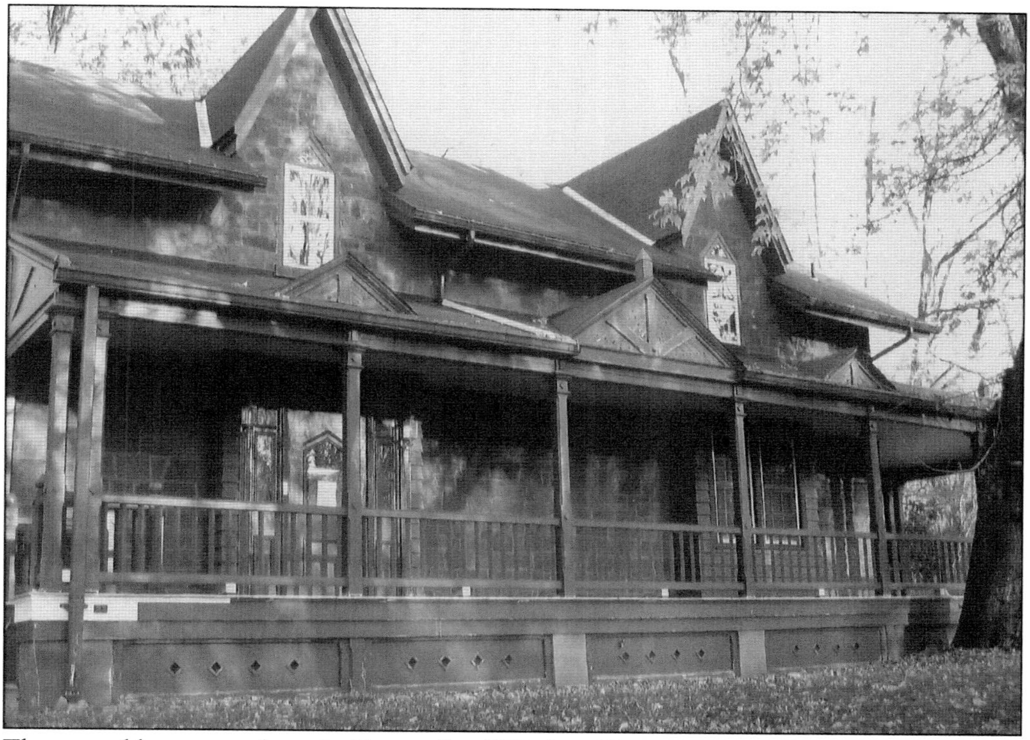

The second-largest employer in Cornwall is PRL, Inc. The corporate office (above) is located adjacent to the Cornwall Iron Furnace in a beautifully remodeled stone building overlooking the Bethlehem Steel machine shop. The old machine shop (below) has been transformed into a very modern machine shop used by PRL, Inc. That facility now employs 95 people and specializes in machining very large metal components of up to 10,000 pounds. Its biggest customers are companies in the nuclear energy field, those involved in the defense industries, and other industrial facilities requiring large, precisely machined parts.

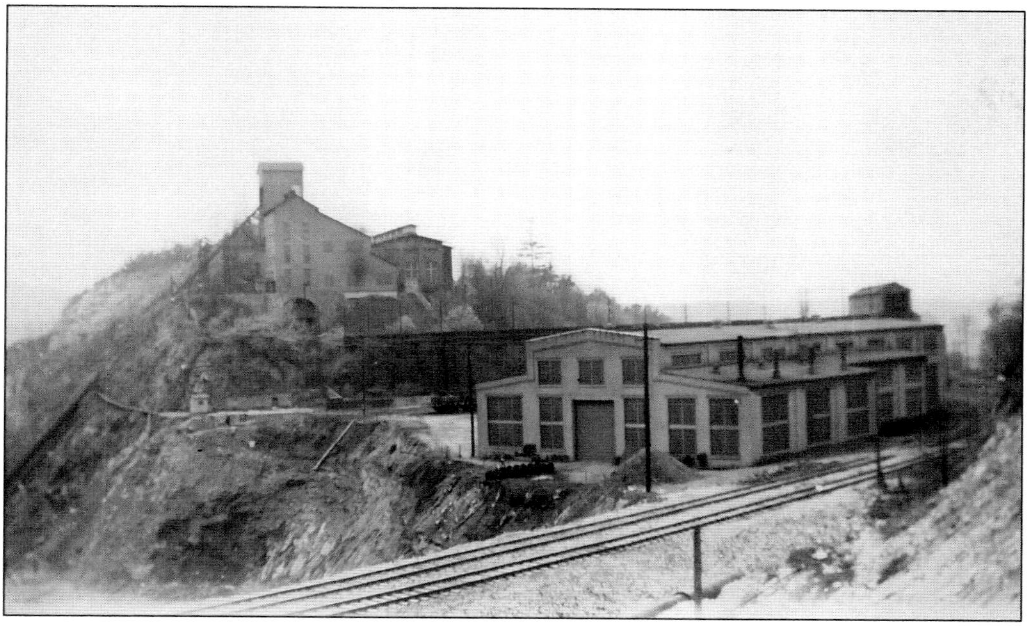

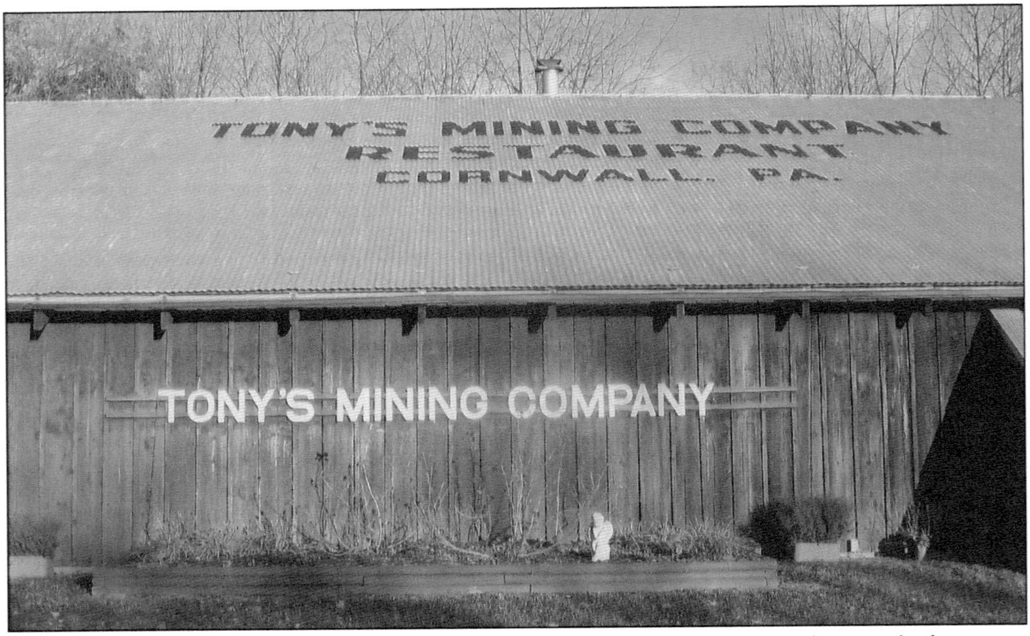

In 1974, Anthony "Tony" Cek built Tony's Mining Company Restaurant. Cek was a draftsman in the engineering department of the mines. His father, John Cek, worked in a mine in Minnesota before coming to Cornwall. Every part of the structure—even the entranceway—was designed to resemble a mine. The restaurant is known for elegant dining, as shown in the image below. (Below, courtesy of Jean Kotkas.)

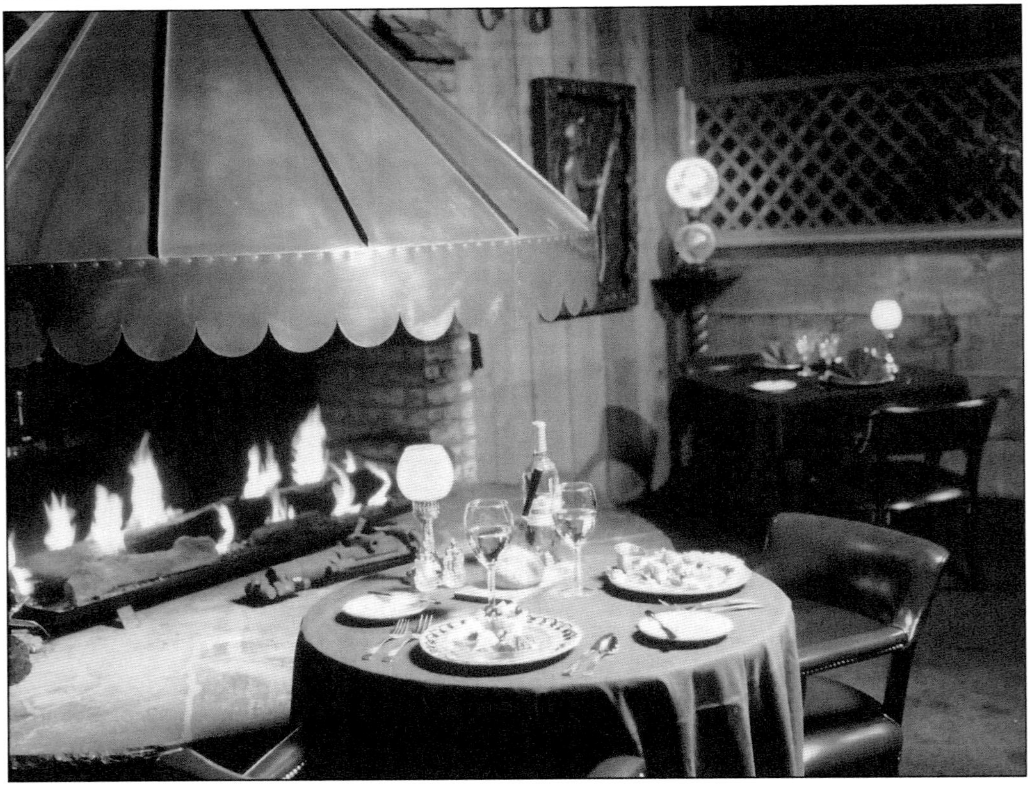

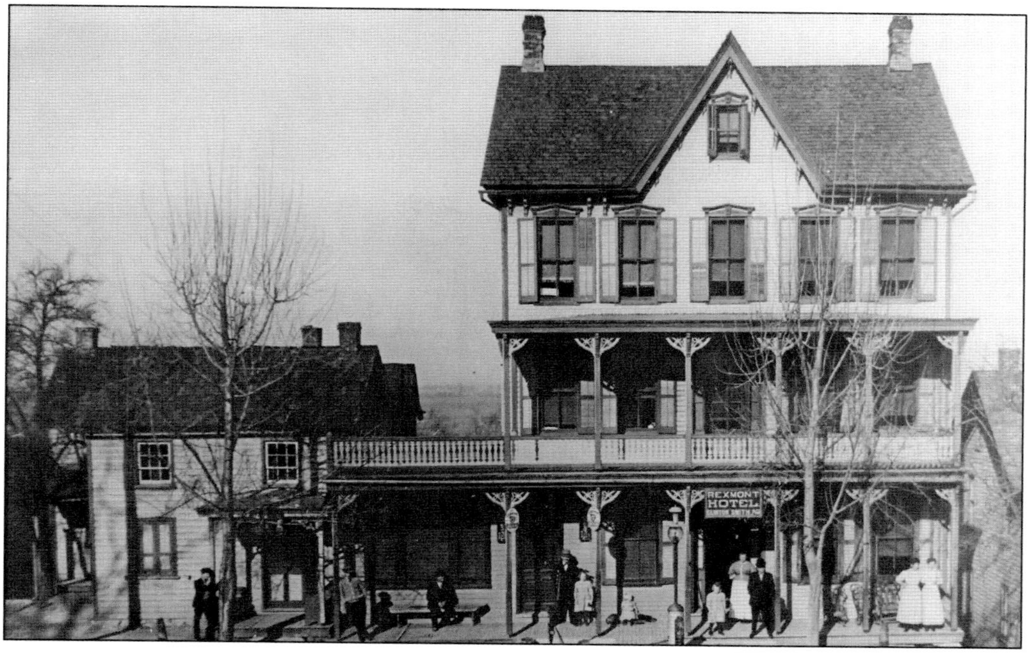

Above is the Rexmont Hotel in 1912. The building was constructed in 1890 by Fredrick Carpenter, and at the time of this picture, it was owned by Clinton Smith. Below is the inside of Boyer's Café in 2010, with Robert Boyer, the owner, at the tap. (Both, courtesy of Robert Boyer.)

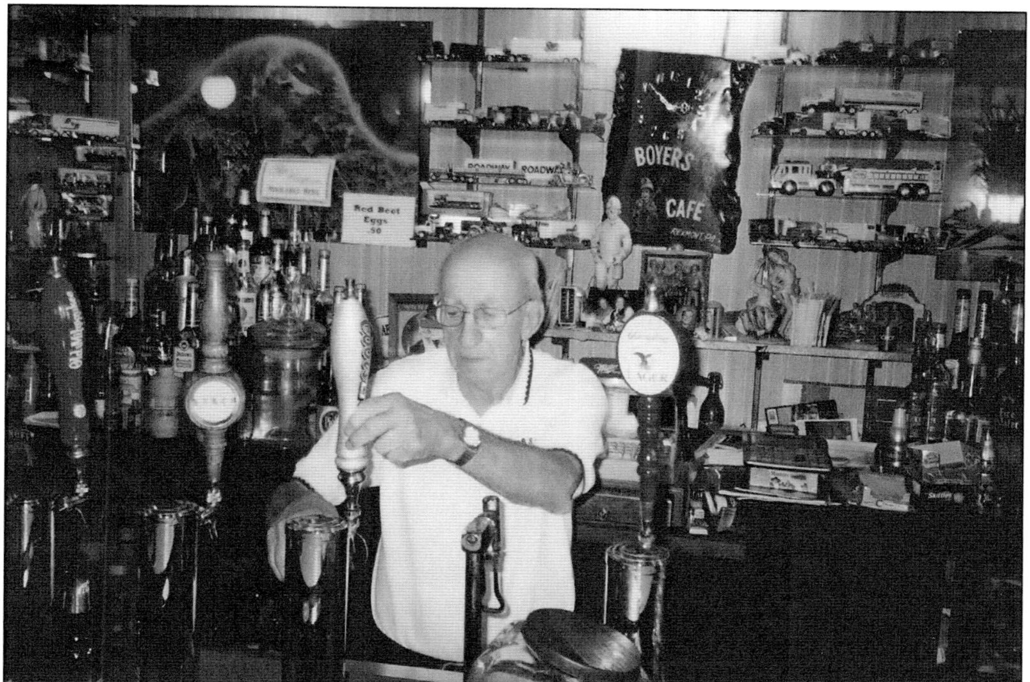

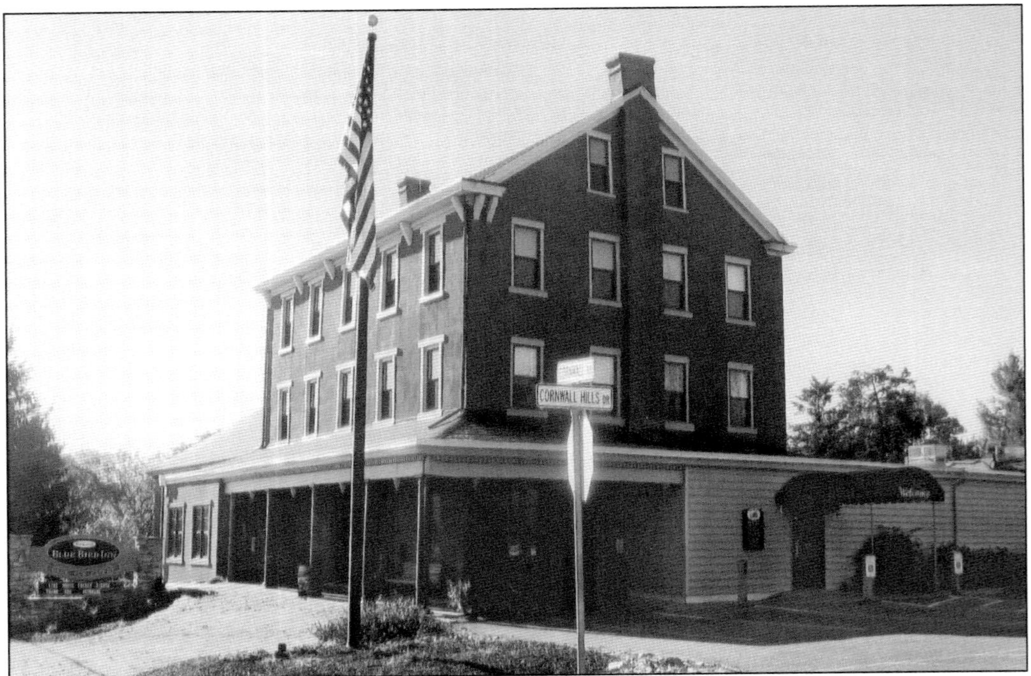

In 1859, Ye Olde Golden Key Hotel was built. The lodging operated under a number of other names until the 1930s, when it became the Blue Bird Inn. The structure has been remodeled a number of times. The most recent updates and addition are shown in the photograph above. The remodeled inside is shown below. The building is located along the Cornwall Pike (formerly the old plank road) a short distance from Cornwall Center. It is the largest and most popular eatery in Cornwall.

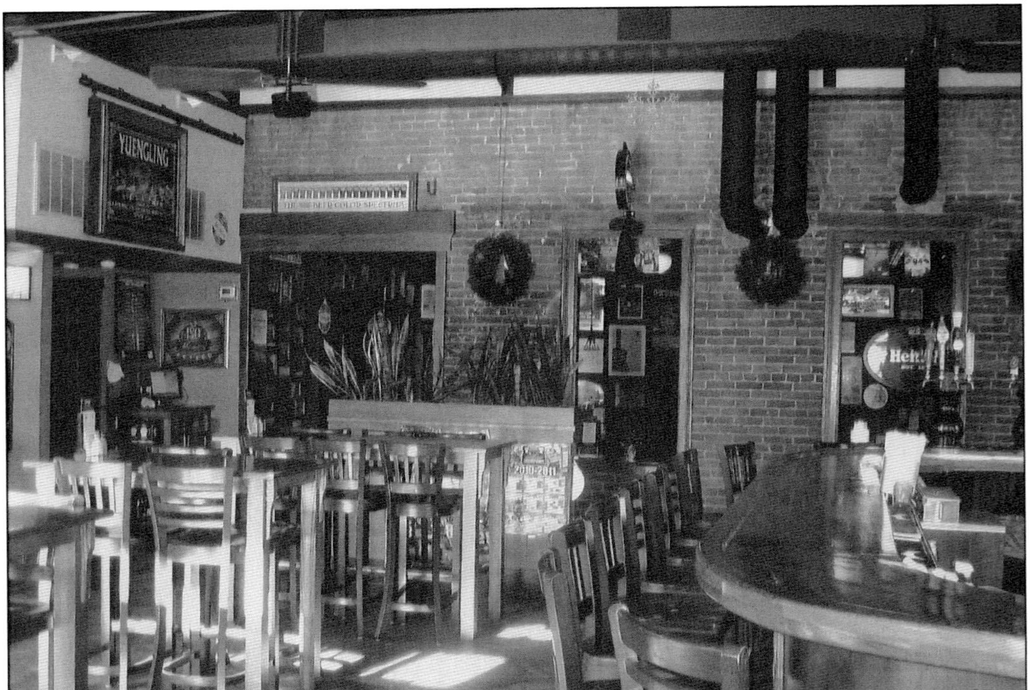

Six

It Is All Underground Now

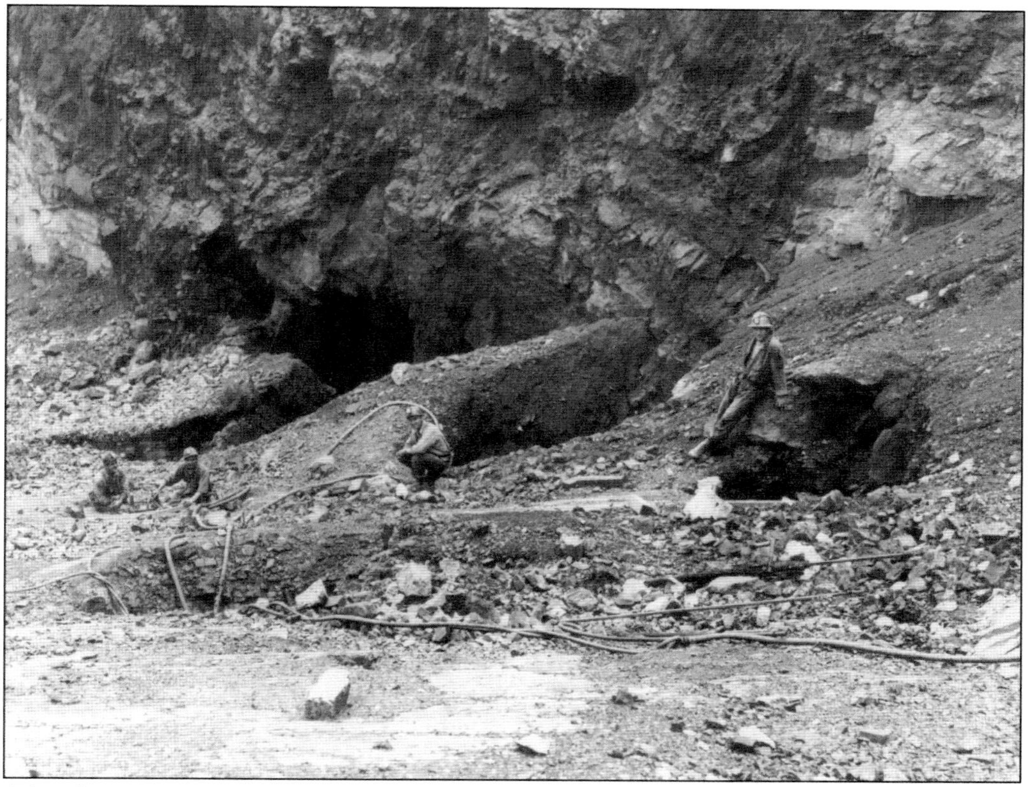

After the open pit was closed, all mining was concentrated underground in the No. 3 and No. 4 Mines. The No. 3 Mine focused on working the ore body around the open pit that could not be reached using the open-pit method. Pictured here are four unidentified miners from No. 3 Mine. A close look shows lamps on their hard hats and battery packs on their belts. Almost in the middle of the image is the place where the No. 3 Mine broke through to the open pit. Miners had started their day underground and ventured into the sunlight later for a break.

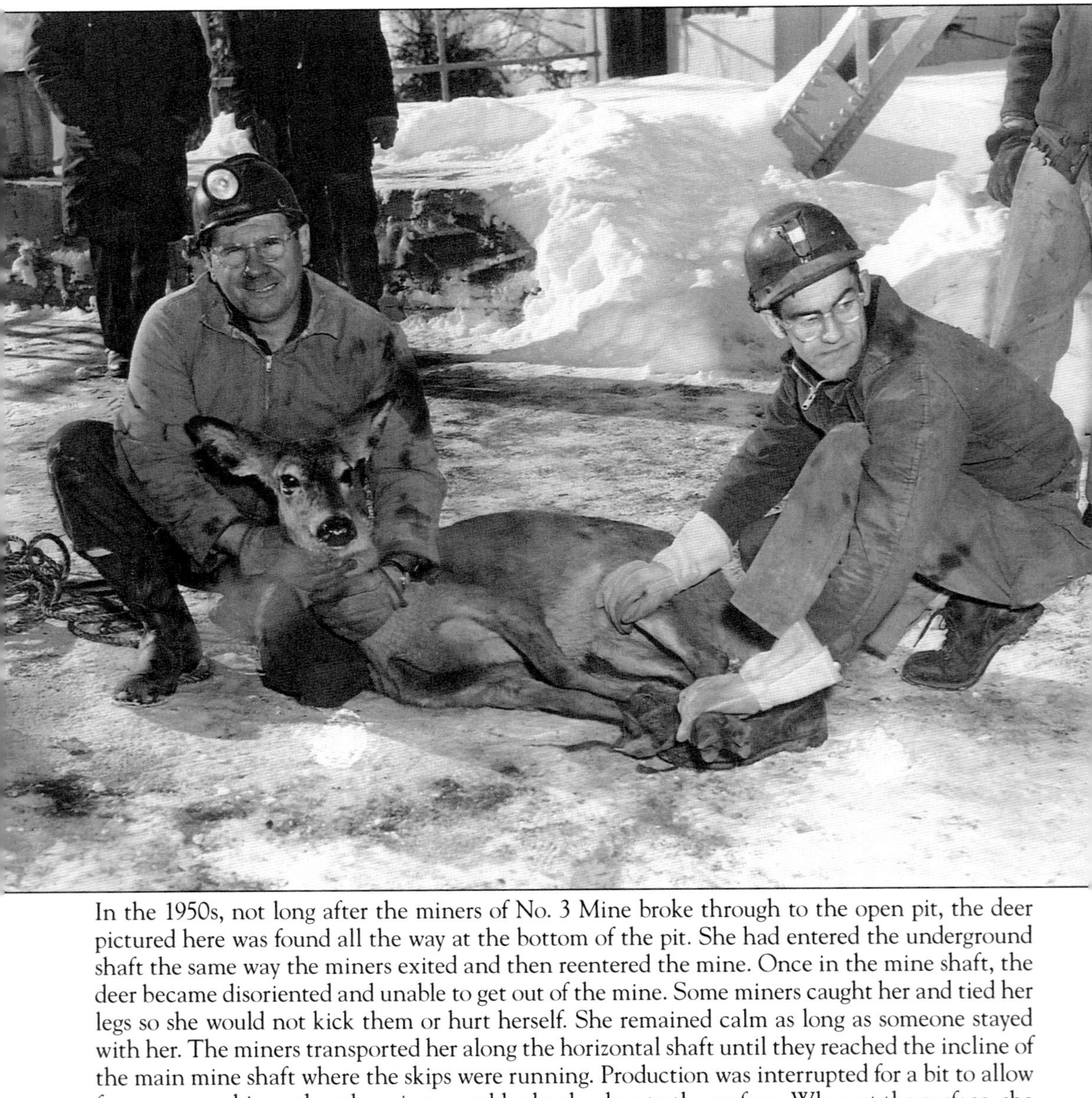

In the 1950s, not long after the miners of No. 3 Mine broke through to the open pit, the deer pictured here was found all the way at the bottom of the pit. She had entered the underground shaft the same way the miners exited and then reentered the mine. Once in the mine shaft, the deer became disoriented and unable to get out of the mine. Some miners caught her and tied her legs so she would not kick them or hurt herself. She remained calm as long as someone stayed with her. The miners transported her along the horizontal shaft until they reached the incline of the main mine shaft where the skips were running. Production was interrupted for a bit to allow for an empty skip so that the miners could take the deer to the surface. When at the surface, she was unloaded from the skip, checked for injury, and released back to the wild.

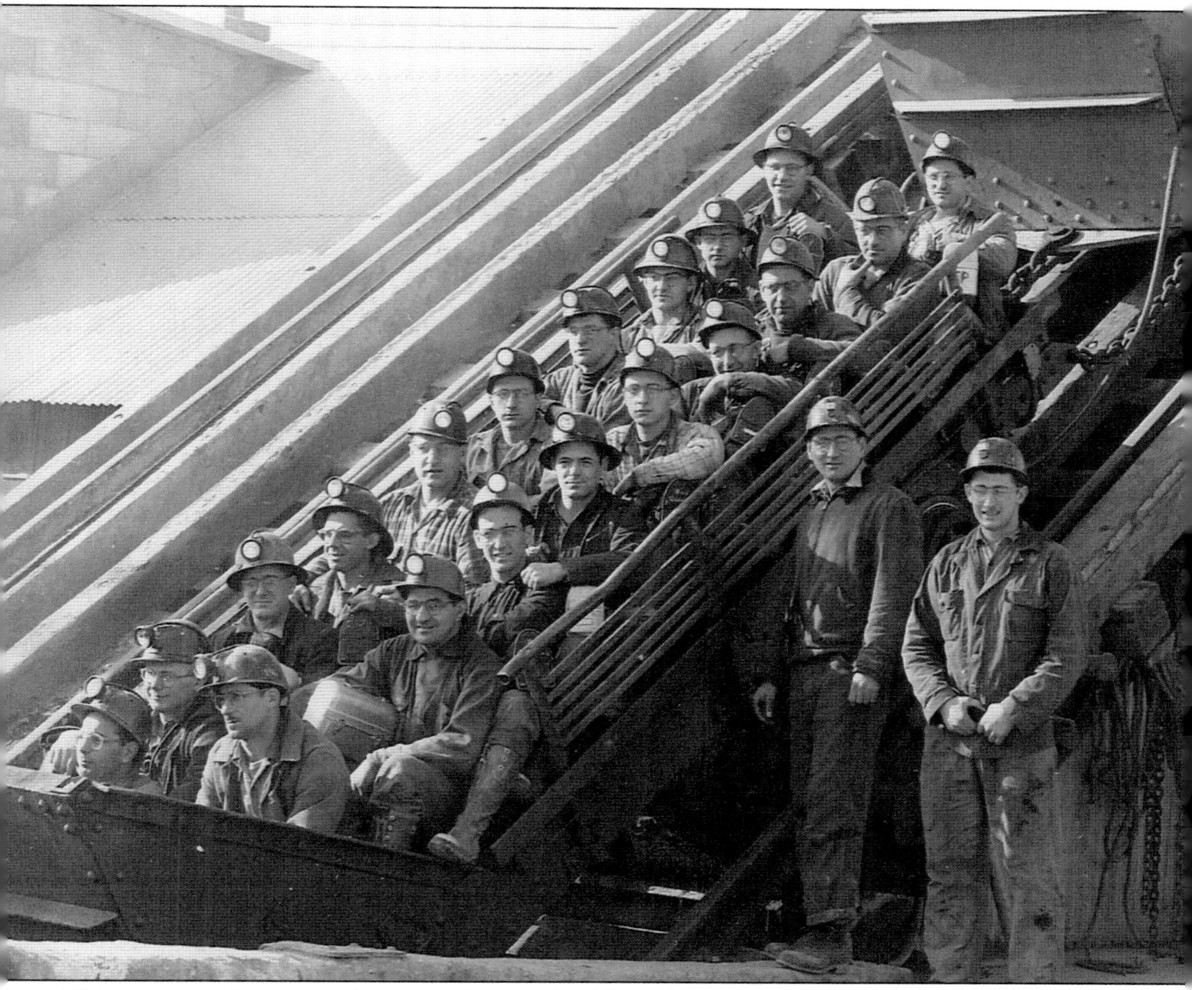

This is a group of miners ready to go underground at No. 4 Mine. The incline slope is 29 degrees—the two men standing beside the skip help to show just how steep the incline leading into the mine below is. The next stop will be 1,200 feet below.

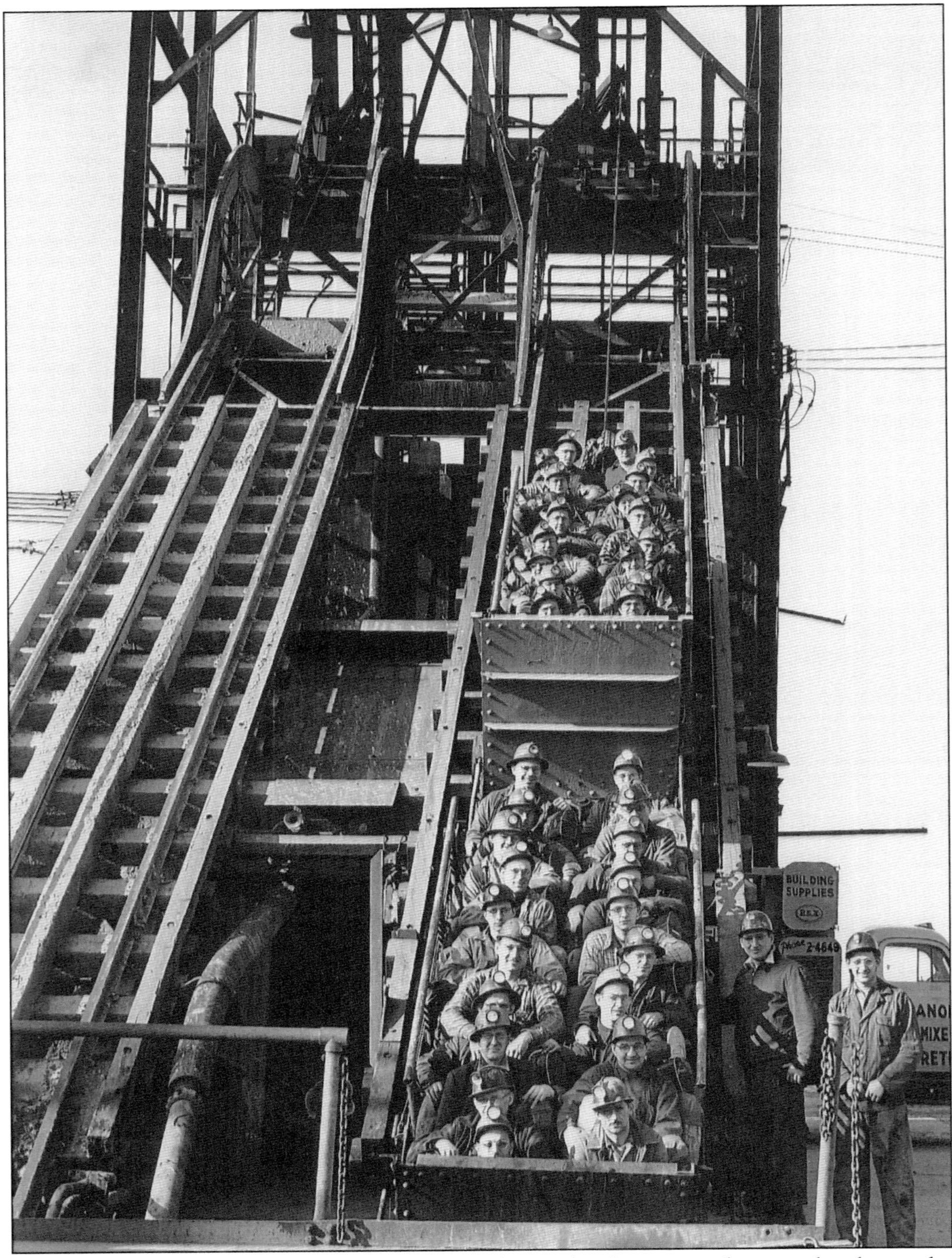

Both skips are fully loaded with miners and ready to take the men underground to begin the workday. This was the transportation in and out of the mine.

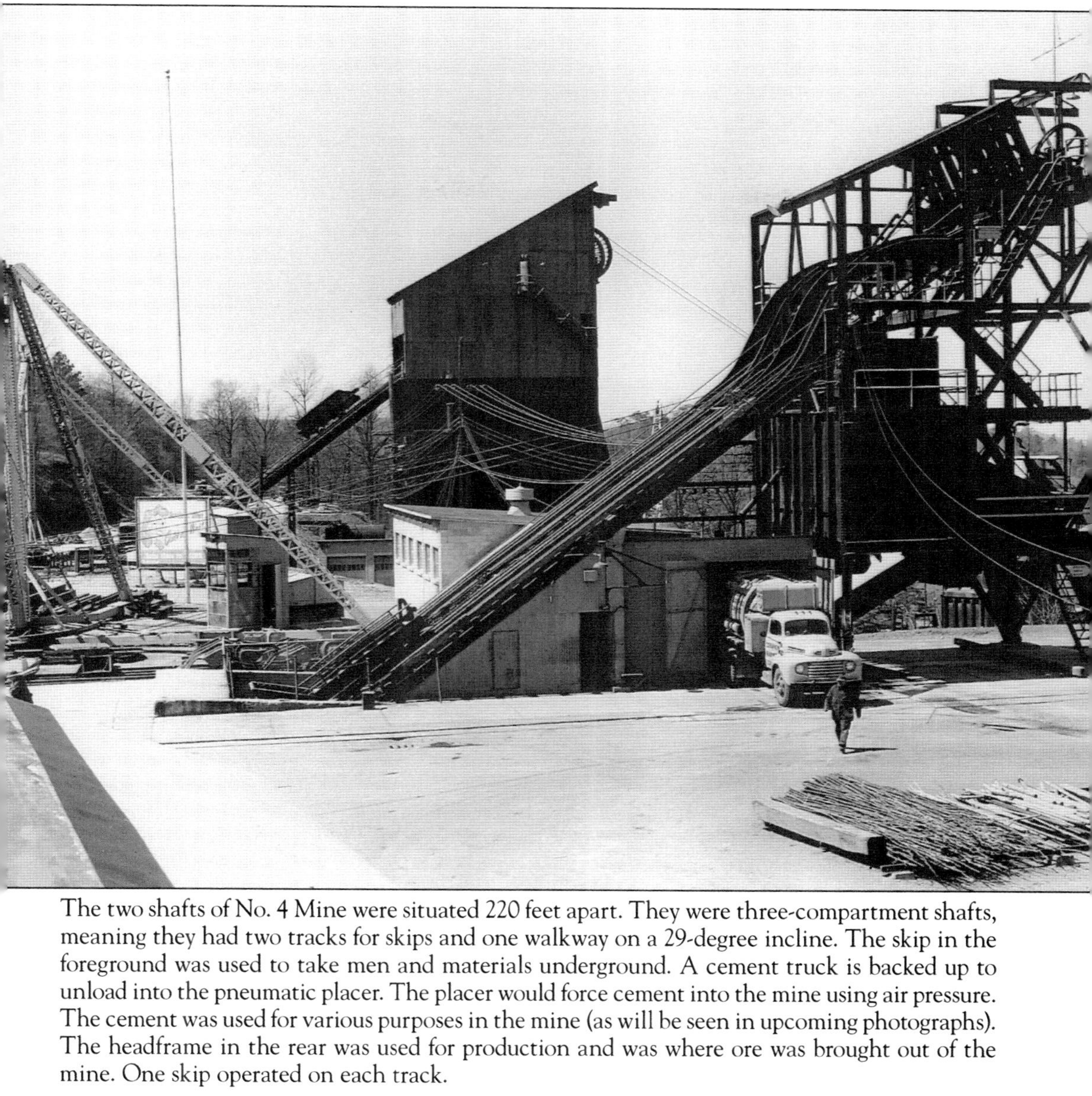

The two shafts of No. 4 Mine were situated 220 feet apart. They were three-compartment shafts, meaning they had two tracks for skips and one walkway on a 29-degree incline. The skip in the foreground was used to take men and materials underground. A cement truck is backed up to unload into the pneumatic placer. The placer would force cement into the mine using air pressure. The cement was used for various purposes in the mine (as will be seen in upcoming photographs). The headframe in the rear was used for production and was where ore was brought out of the mine. One skip operated on each track.

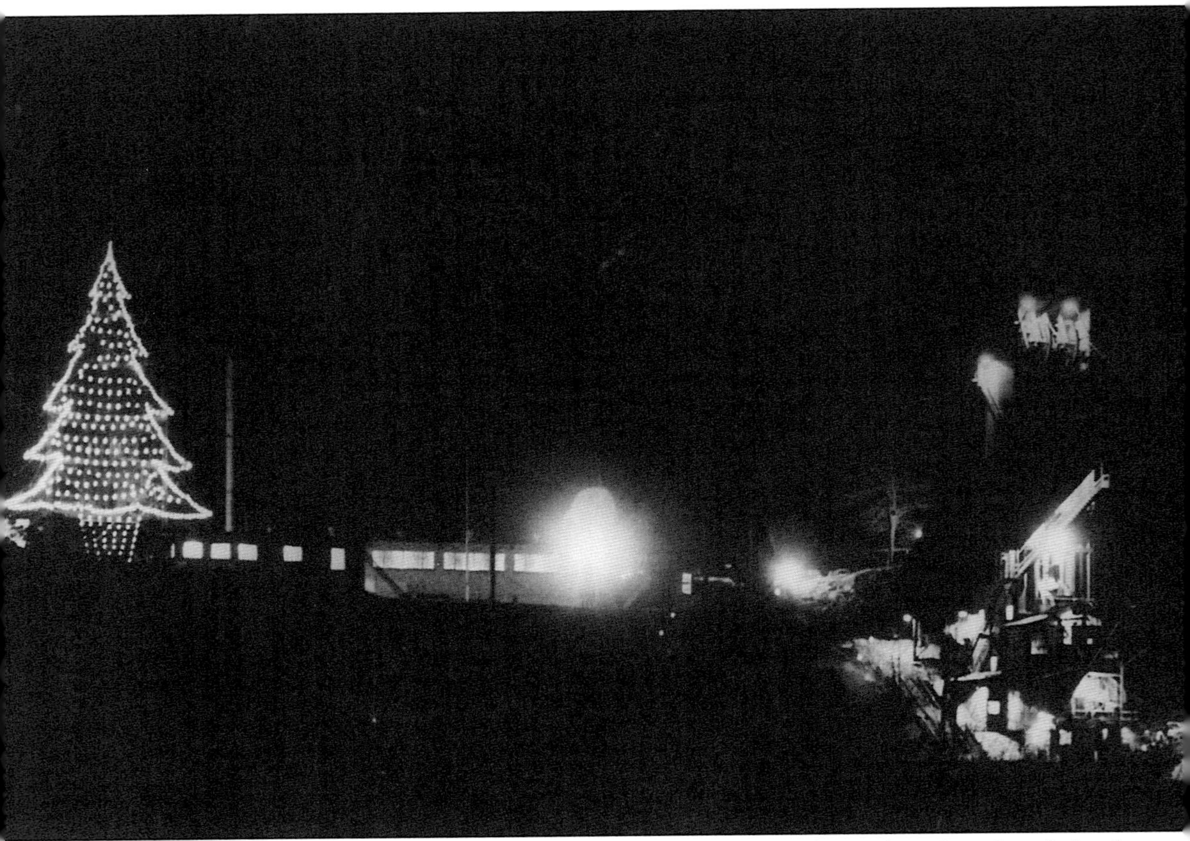

No. 4 Mine's 90-foot-tall headframe (also known as a tipple) was the perfect place for a lighted Christmas tree. Because the headframe was located on top of a hill, the Christmas tree could be seen for many miles when illuminated. The tree was lit in green lights, and the stand for the tree was lit in red. The No. 3 Mine had a large star made of all-white lights fastened to its 90-foot-tall headframe. The No. 3 headframe was not on a hill, so the star was not as visible as this tree.

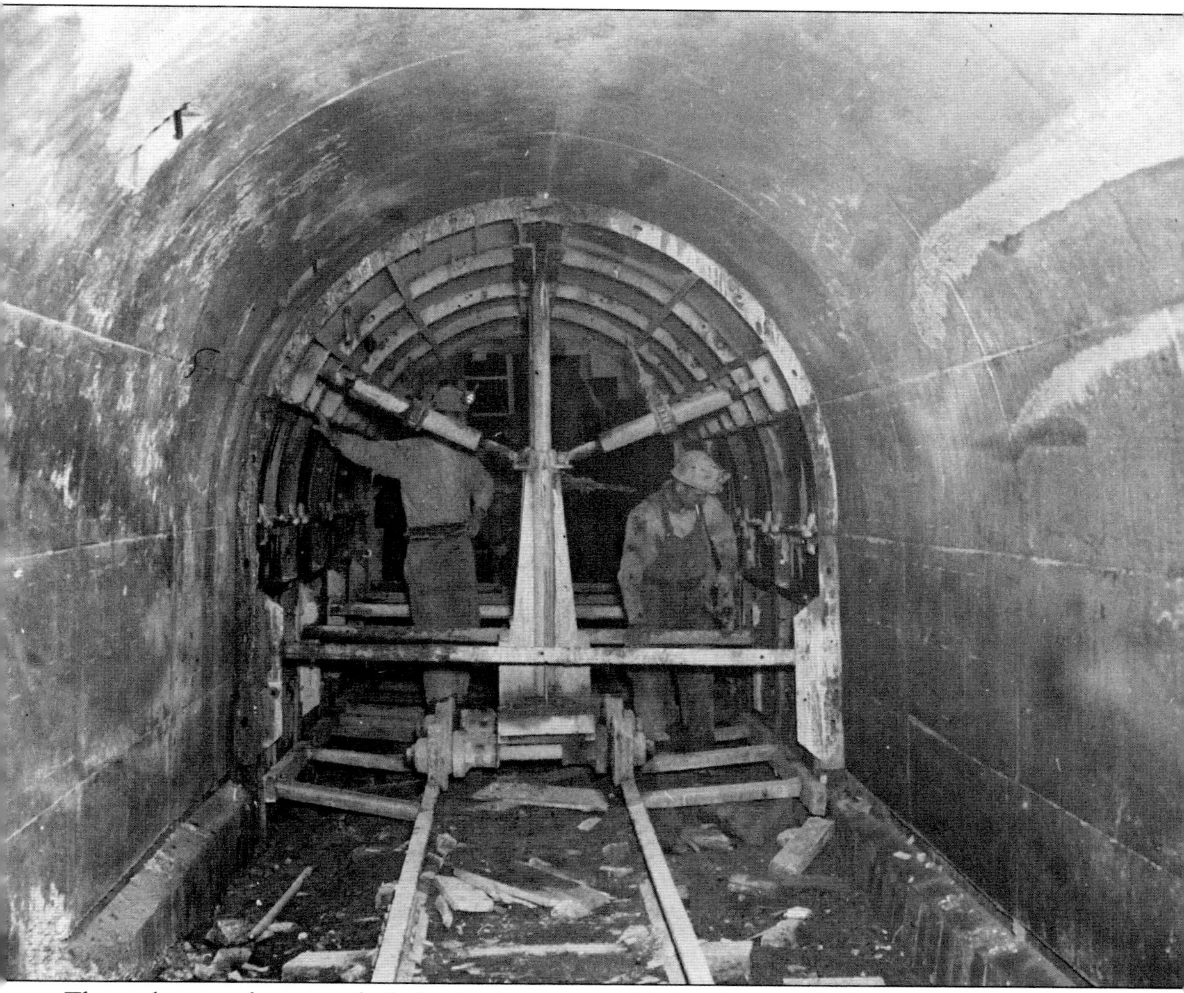

The underground mining shafts had to be concreted. This photograph shows the 10-foot-long form used in that process. It was moved on rails and set into place. The form created a space for the concrete. An aboveground pneumatic placer used air pressure to blow concrete into the gap between the form and the rock walls.

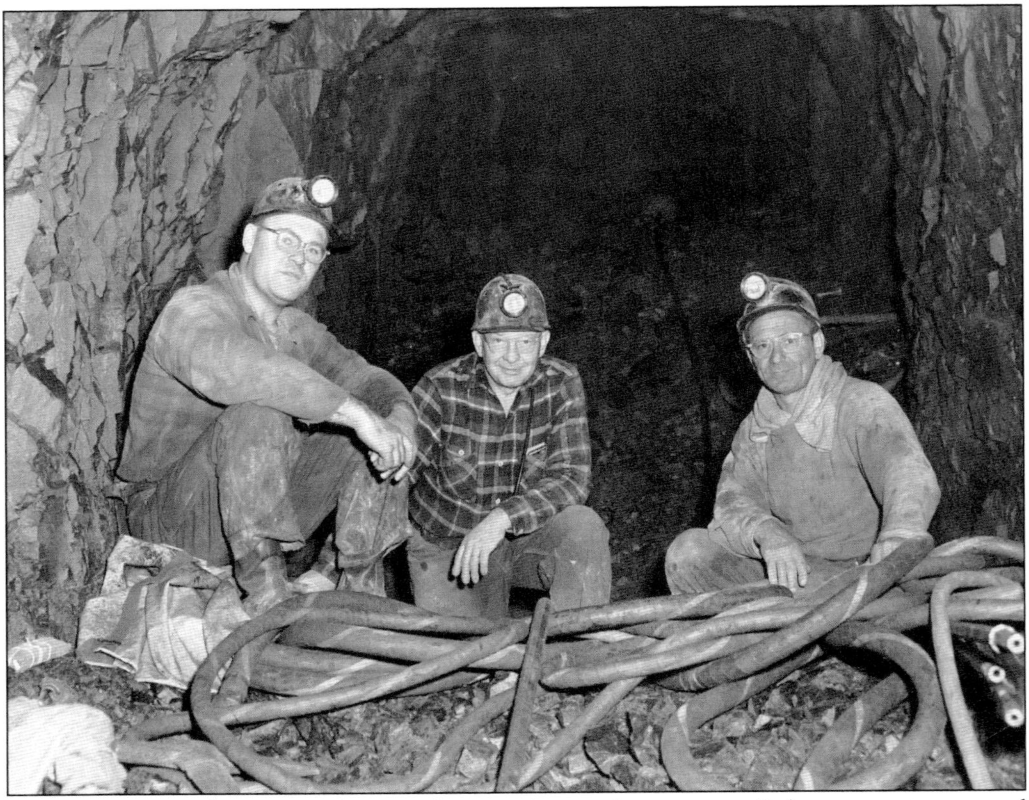

Above, a group of miners working underground poses for a picture. Below, two miners stand by the No. 3 Mine incline while waiting to go back into the mine. The open pit can be seen in the background.

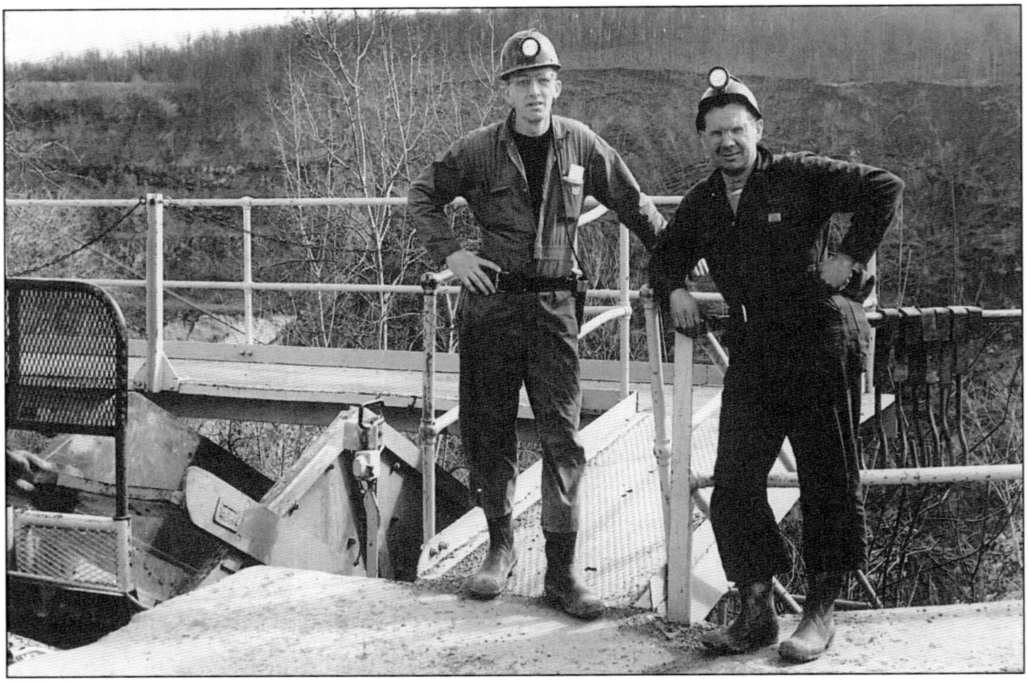

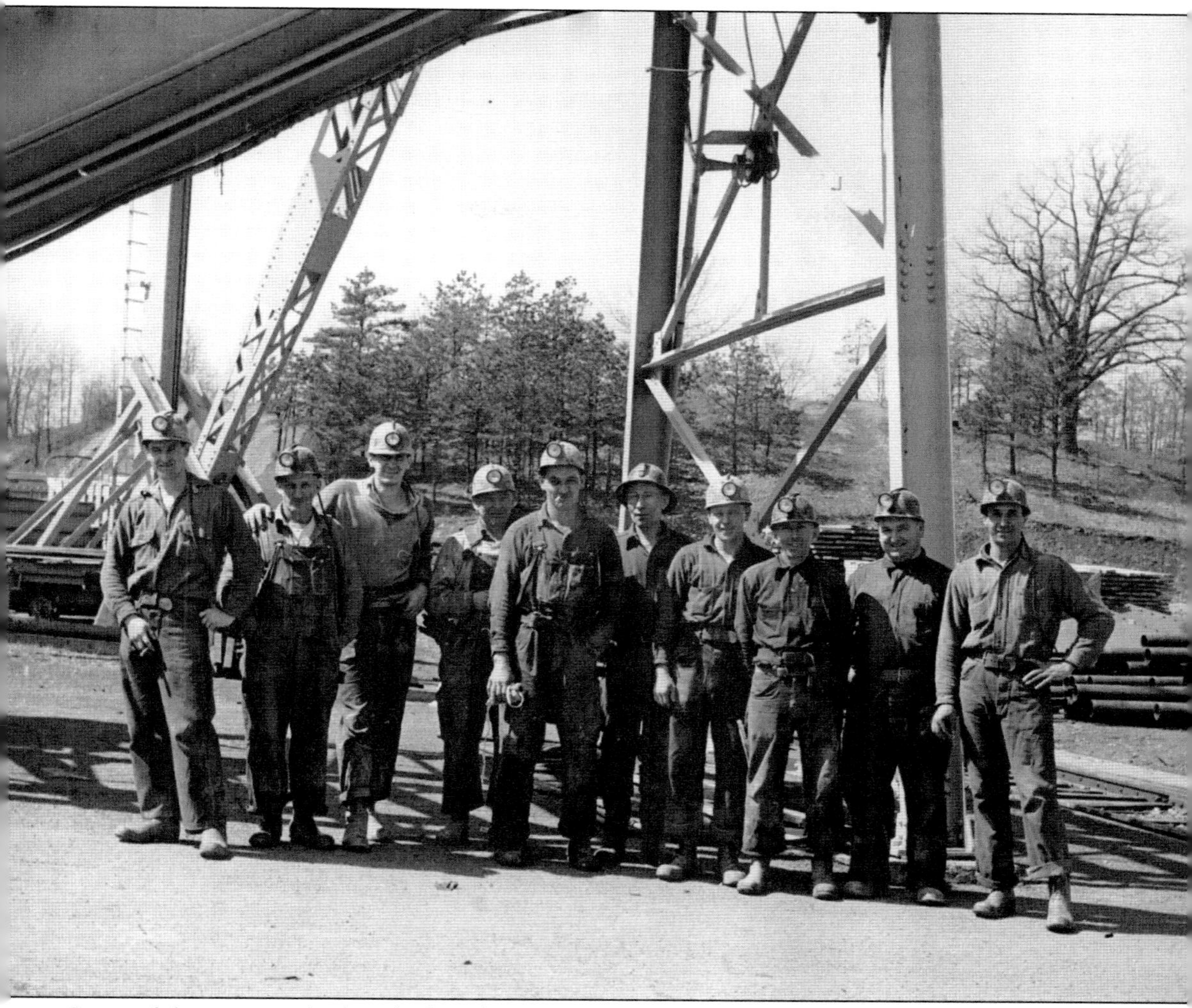

Some of these miners had worked in the open pit before transitioning to underground work at No. 3 Mine. They are shown here getting ready for the start of their shift. A cropped version of this photograph appears on the cover, but the version included here is not cropped and shows the entire group of miners. The angled incline can be seen in the background to the left.

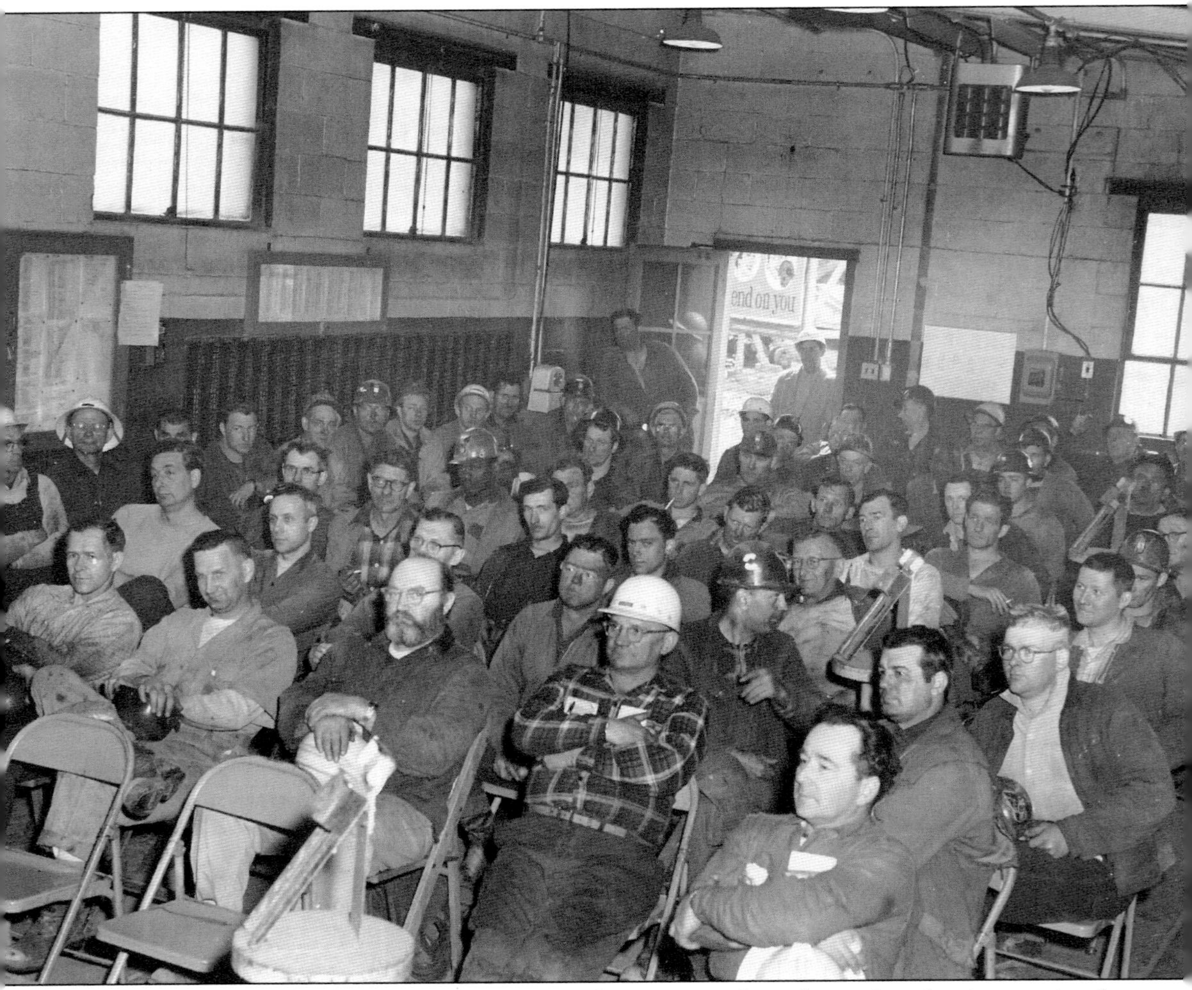

Safety was stressed at the mines. Here, a group of miners attends a safety and production meeting. In 1966, Cornwall No. 4 Mine received the Sentinels of Safety award. It was the highest mining safety award awarded in the United States at that time. The award was jointly given by the American Mining Congress and the US Bureau of Mines. This safety meeting was held aboveground.

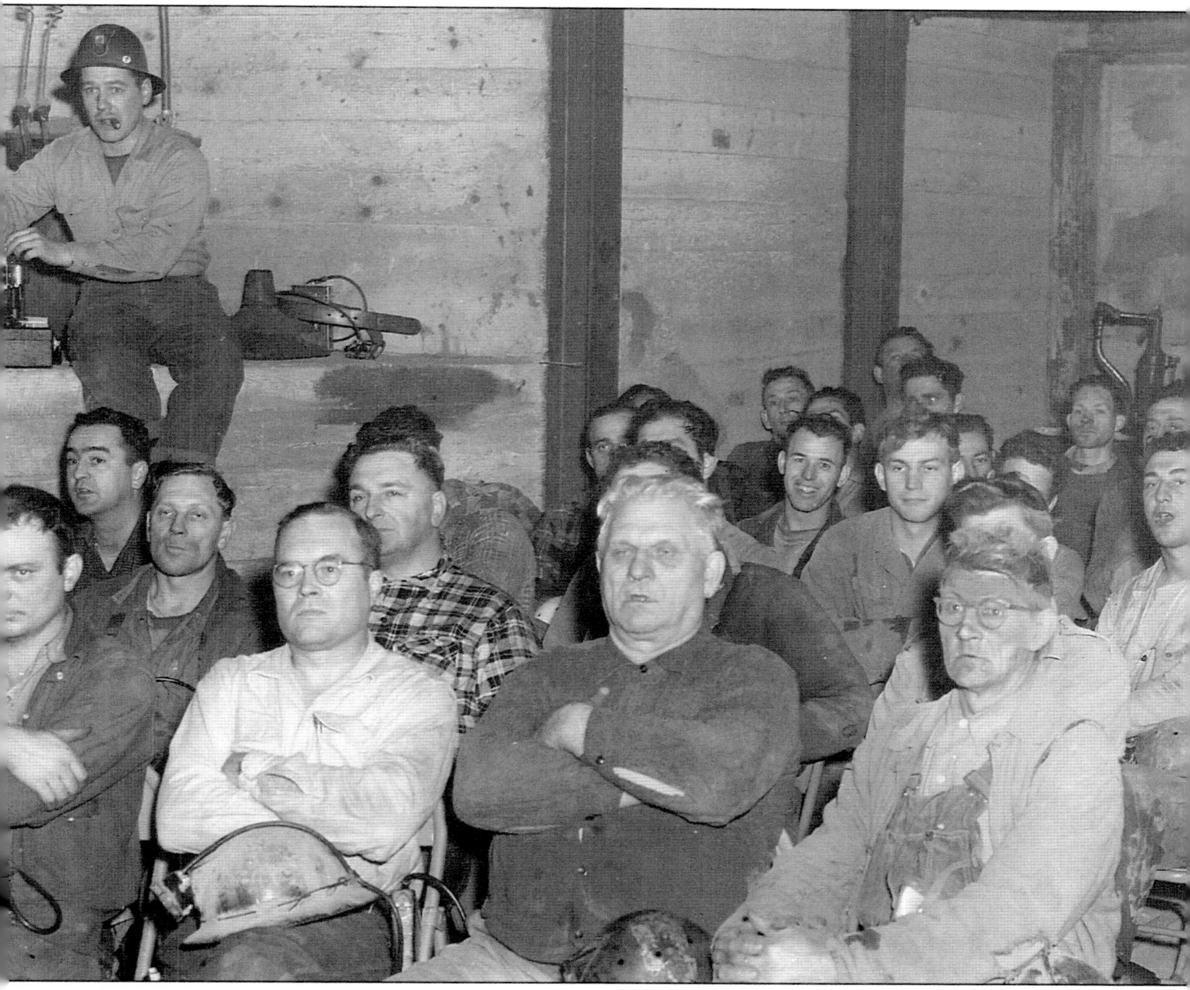

Safety meetings were held aboveground as well as underground. This shift of miners is pictured attending an underground safety meeting.

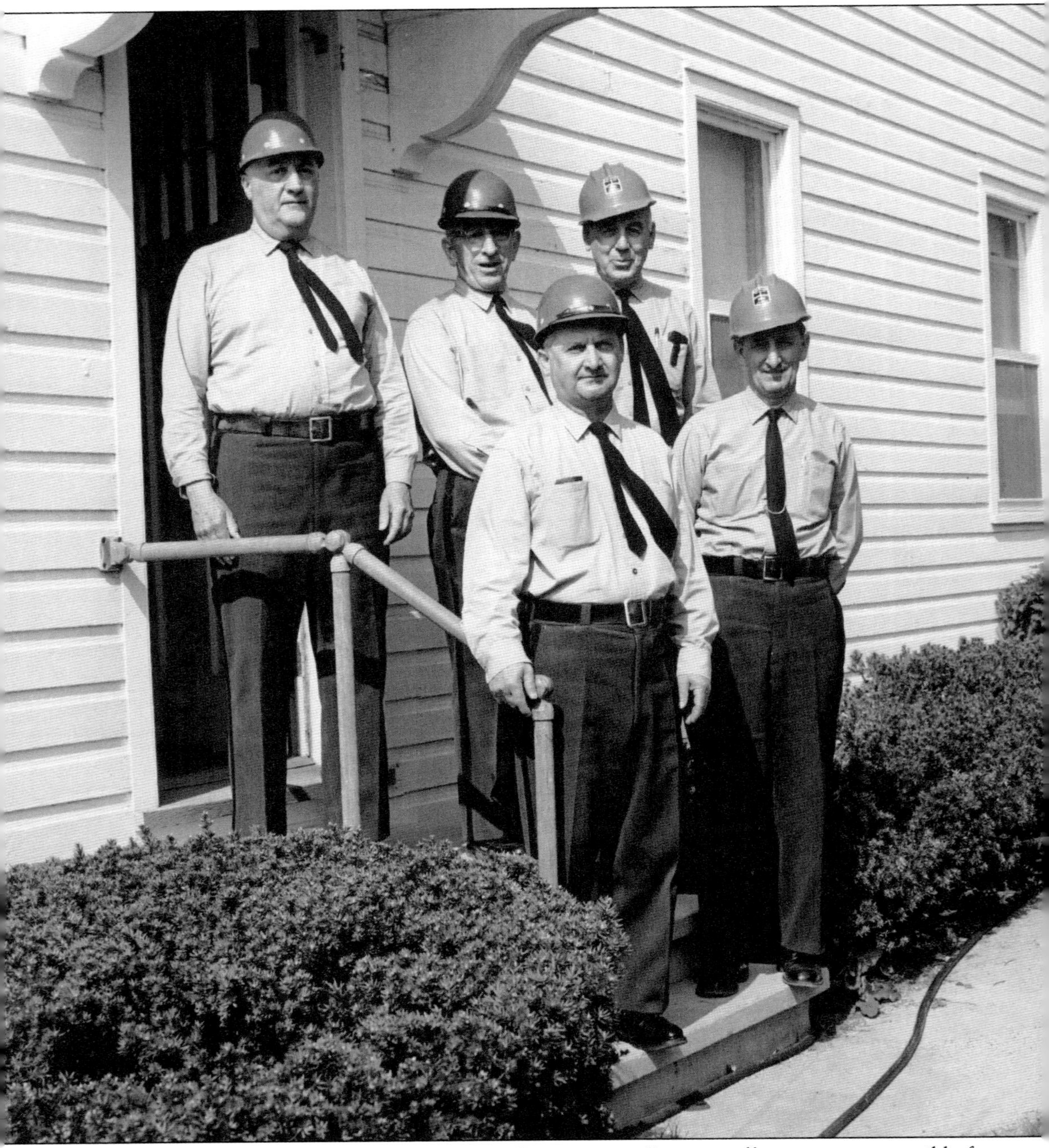

The mining company had a Safety Security Department. Security officers were responsible for making checks of the property, delivering mail between the Bethlehem Steel plants in Cornwall, Lebanon, and Morgantown, and transporting injured employees to the hospital or dispensary. In the first row at far right is Paul Boyer, the last employee of Bethlehem Steel at Cornwall.

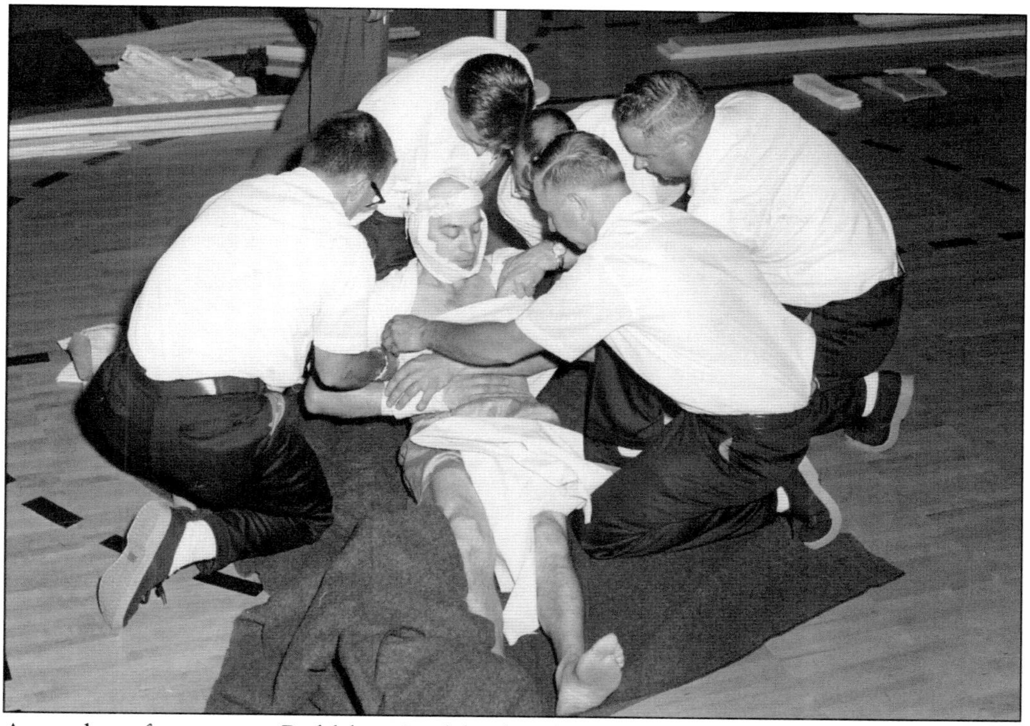

As another safety measure, Bethlehem Steel developed first-aid teams. A number of Bethlehem Steel sites participated in annual first-aid team competitions. The image above shows a team in action administering first aid to a "victim." The image below shows the No. 4 Mine first-aid team.

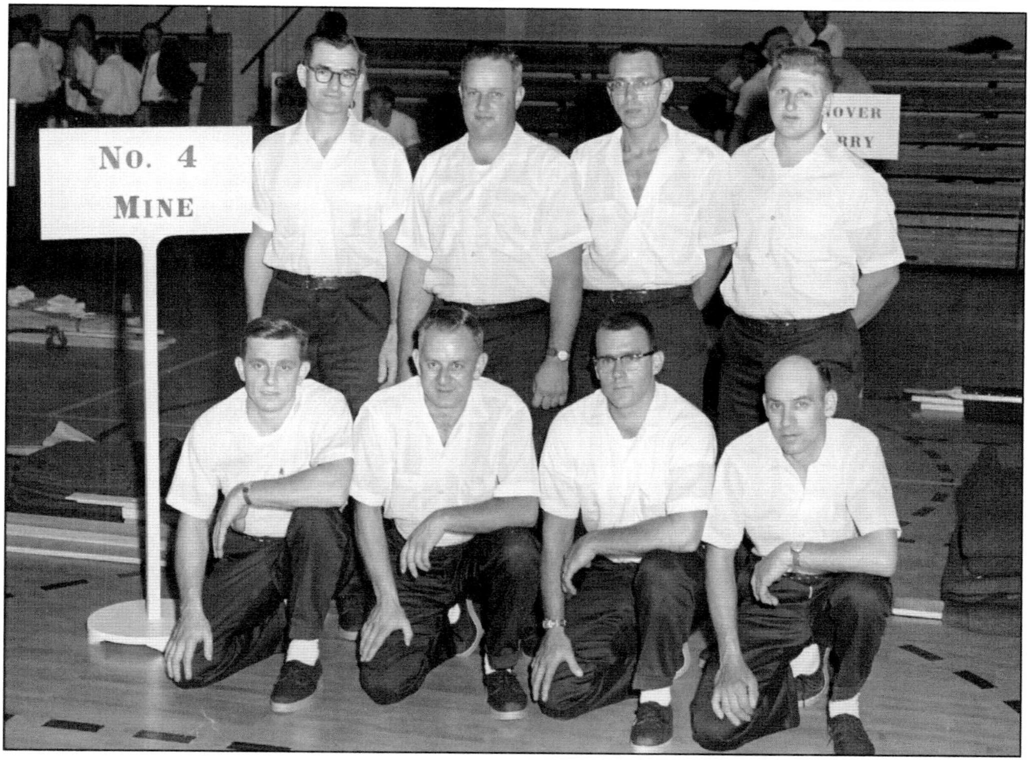

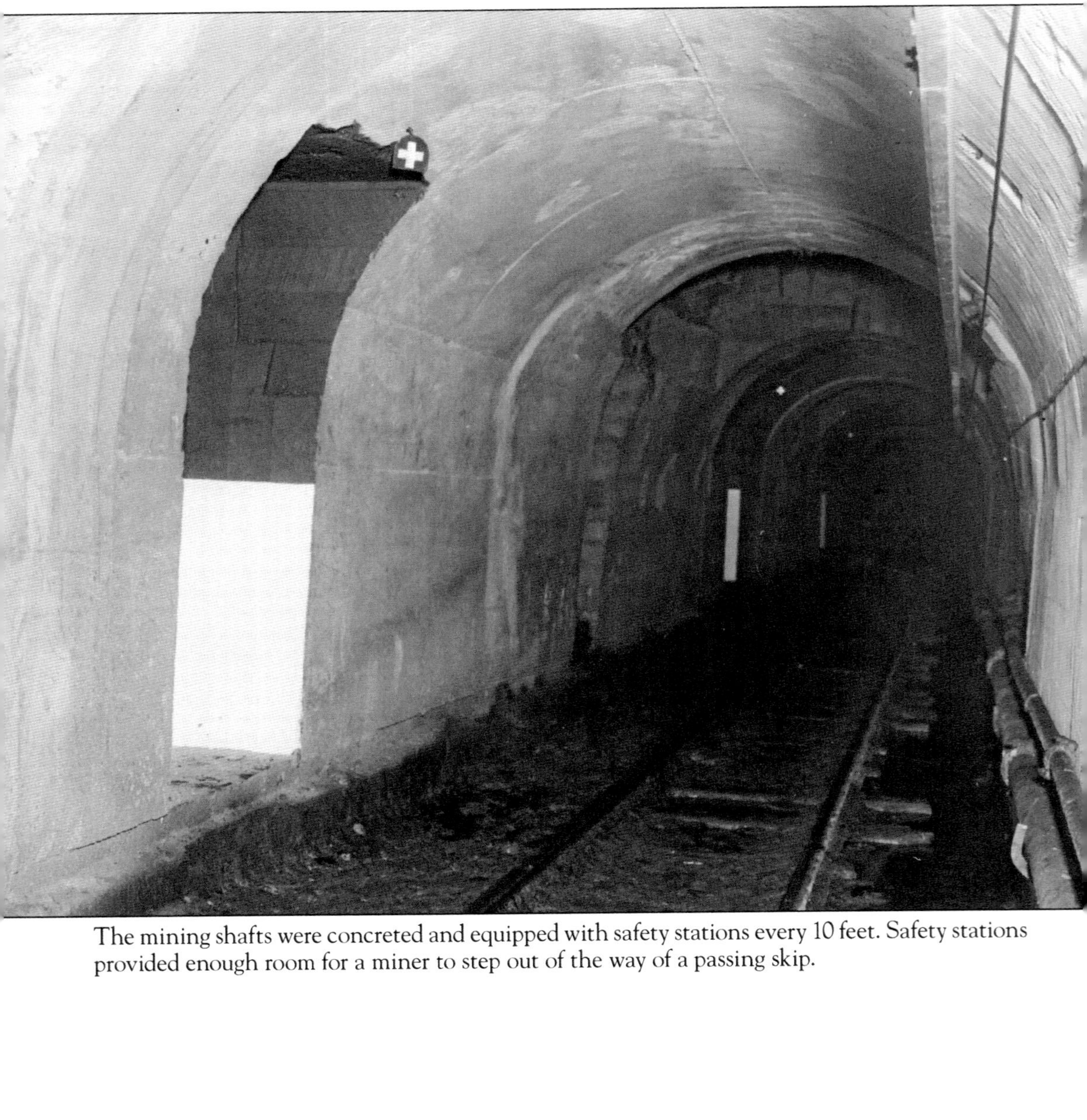

The mining shafts were concreted and equipped with safety stations every 10 feet. Safety stations provided enough room for a miner to step out of the way of a passing skip.

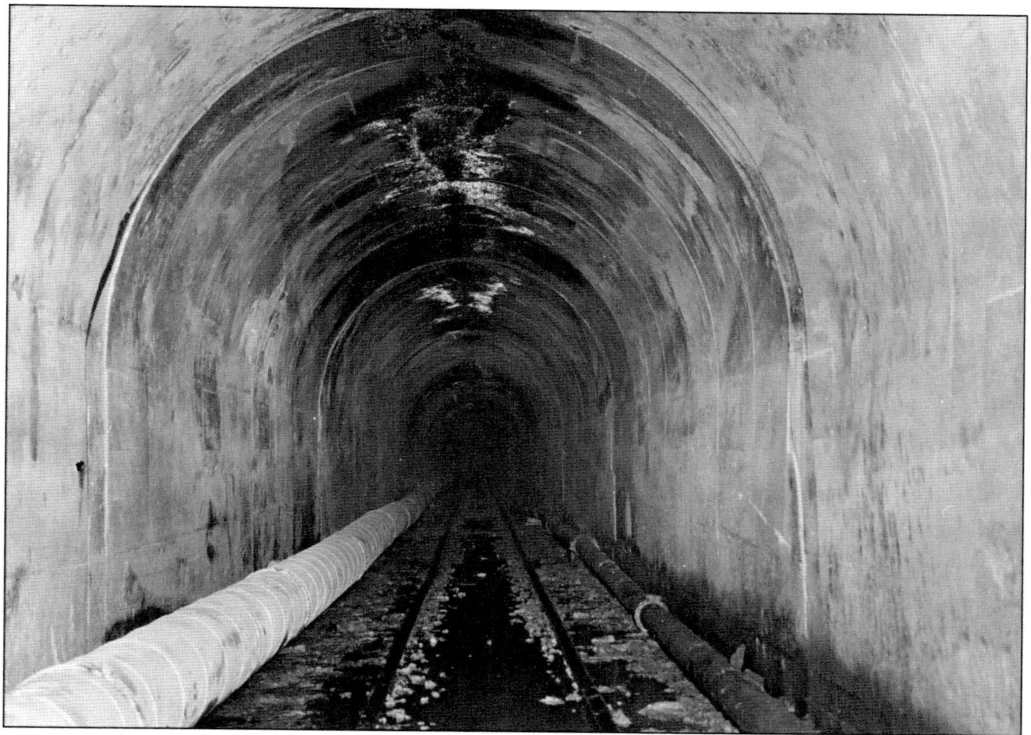

These photographs show a different type of mining shaft known as a horizontal drift. The shafts include neither a headframe nor an incline, and skips would not be used. Horizontal drifts are concreted. The image below shows an intersection with a manual switch to move equipment in one direction or the other. Ore would be moved from these drifts to slots and then loaded on skips to be pulled from the mine.

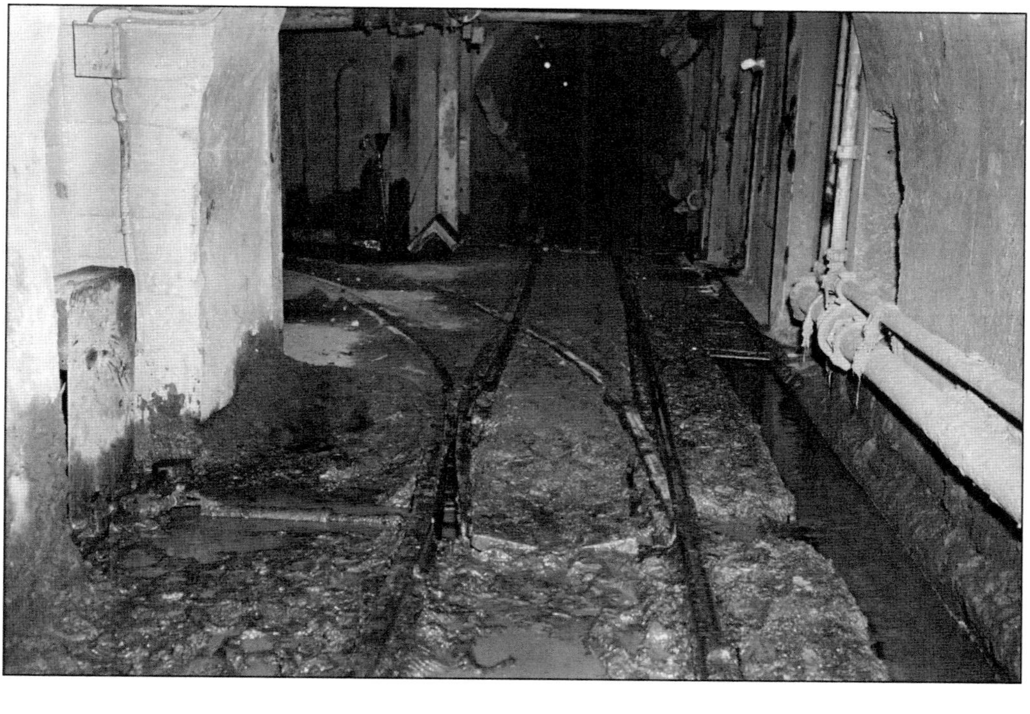

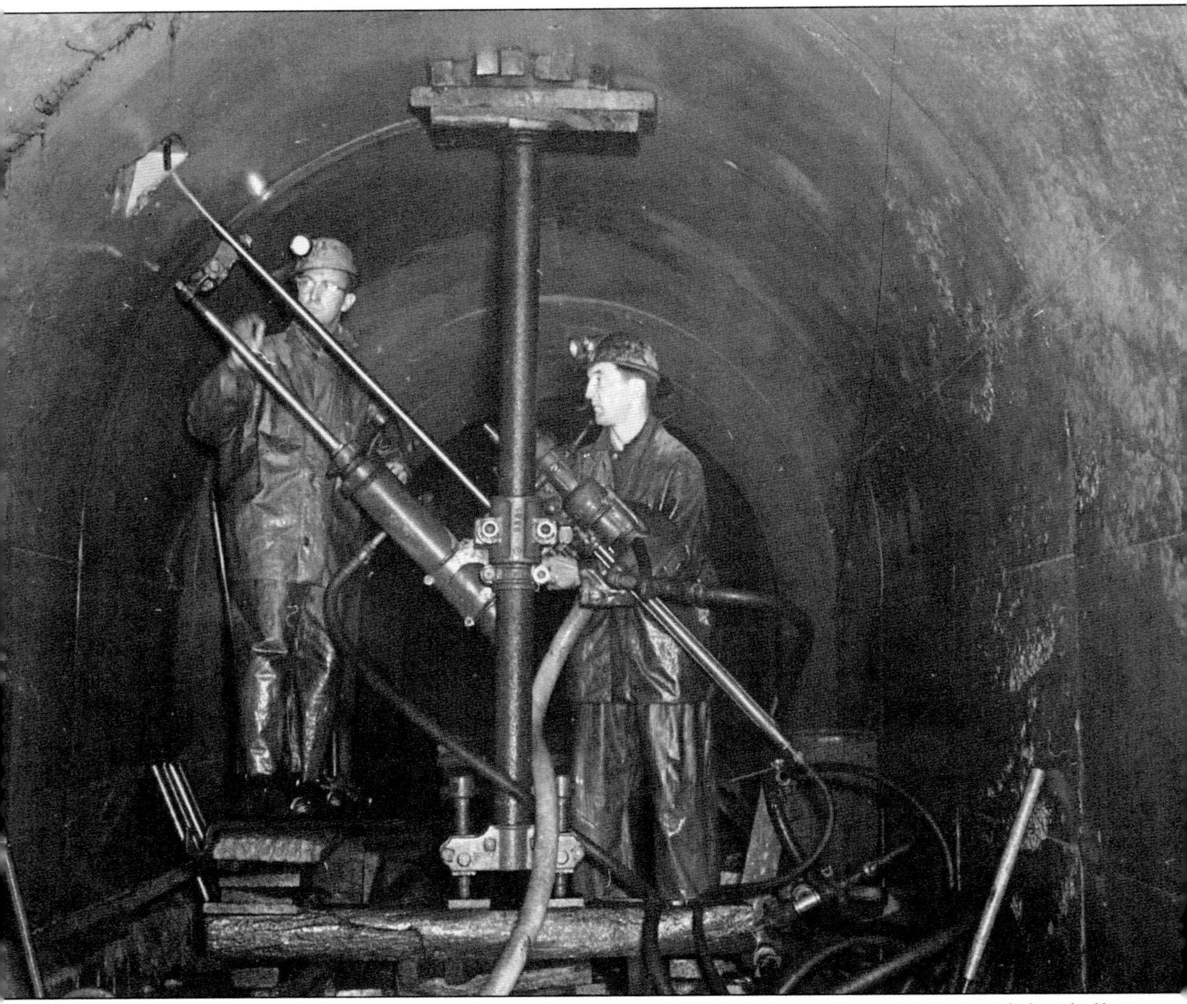

Diamond drills were used underground. Diamonds were embedded into the front of the drill's hollow bit. The hollow section of the bit and the drill would allow a core of rock to be extracted. When the core was removed from the drill, mine engineers could tell the depth of the ore. Here, two unidentified workers operate a diamond drill.

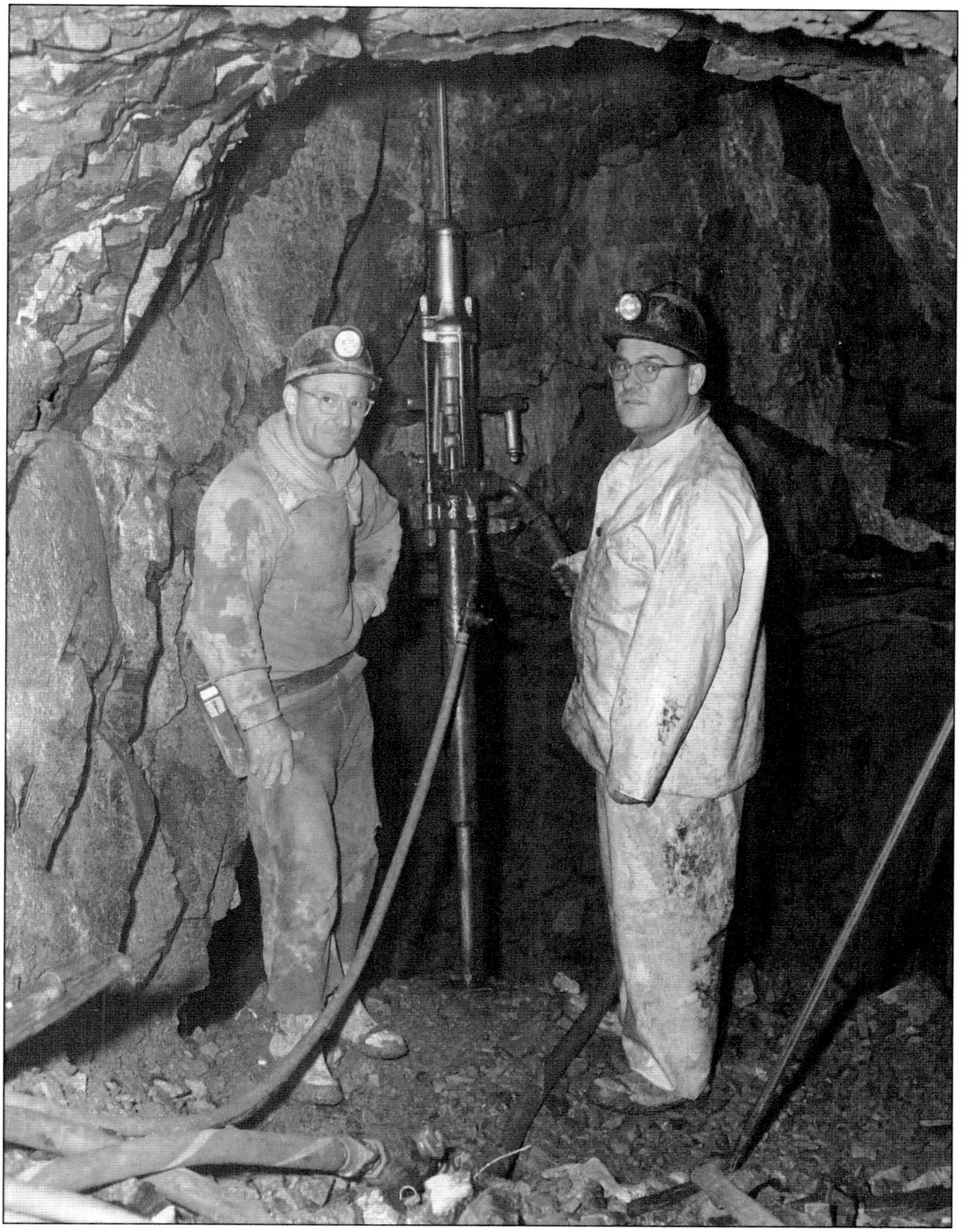

Drills were also used to create the holes for blasting. These two unidentified miners were drilling holes to be loaded with dynamite.

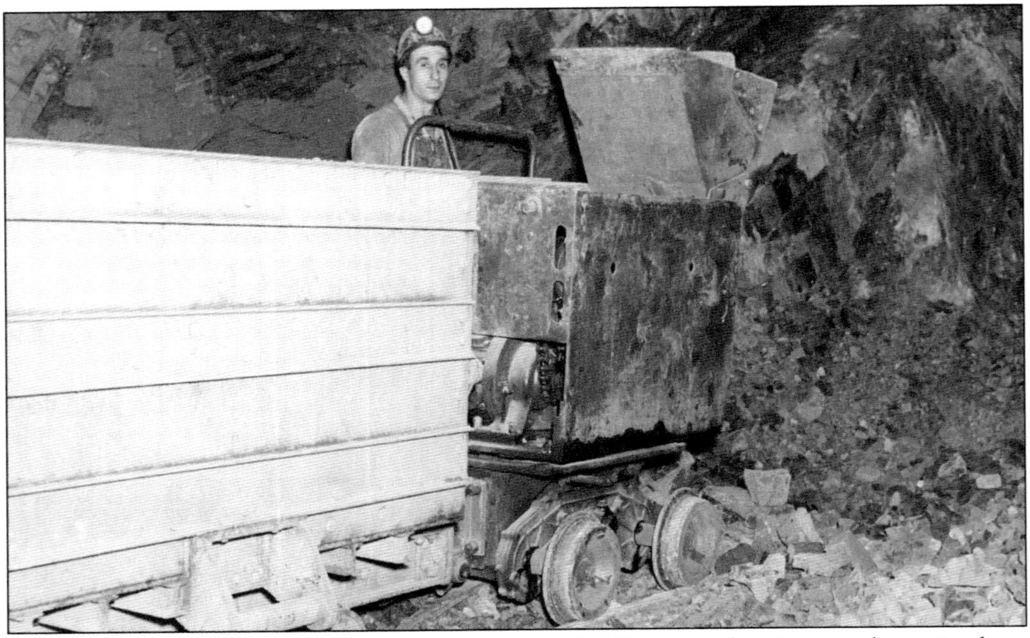

Various types of machinery were used to remove rock. The type of equipment shown in these photographs was known as a "Jitterbug." The picture above shows a small Jitterbug with only one car. The scoop in the front would lift up rock and dump it into the car. In the photograph below is a larger version with multiple cars. The scoop would pick up rock and deposit it on the conveyor belt where the material would be directed into one of the cars.

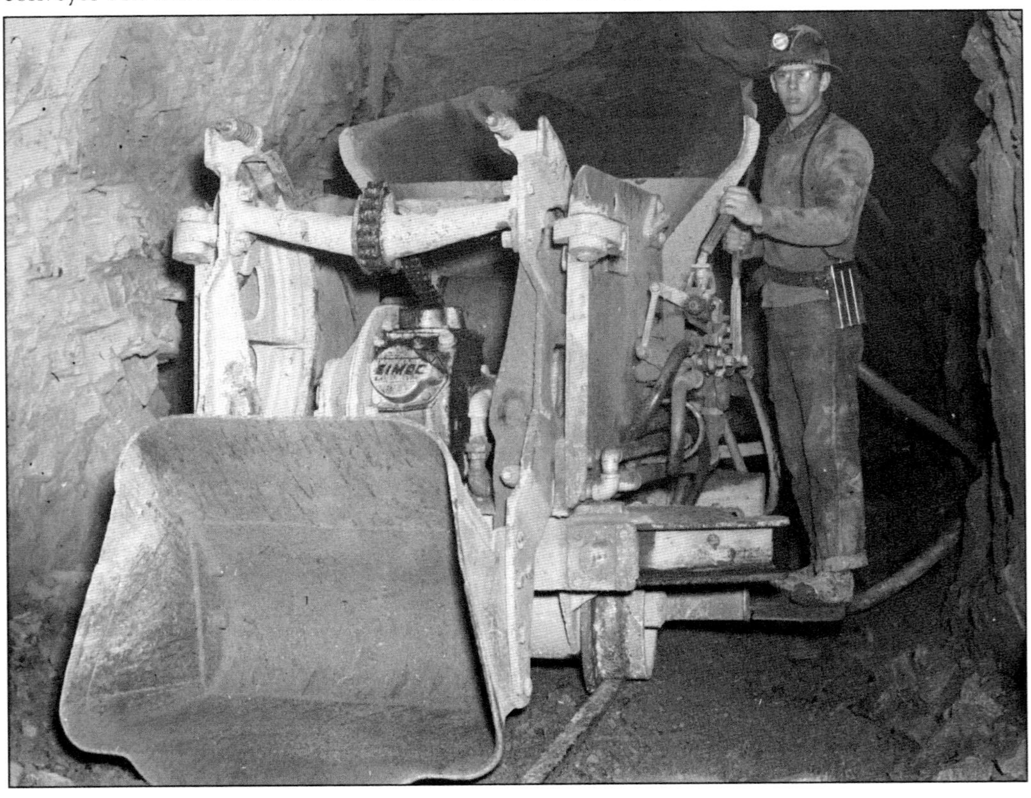

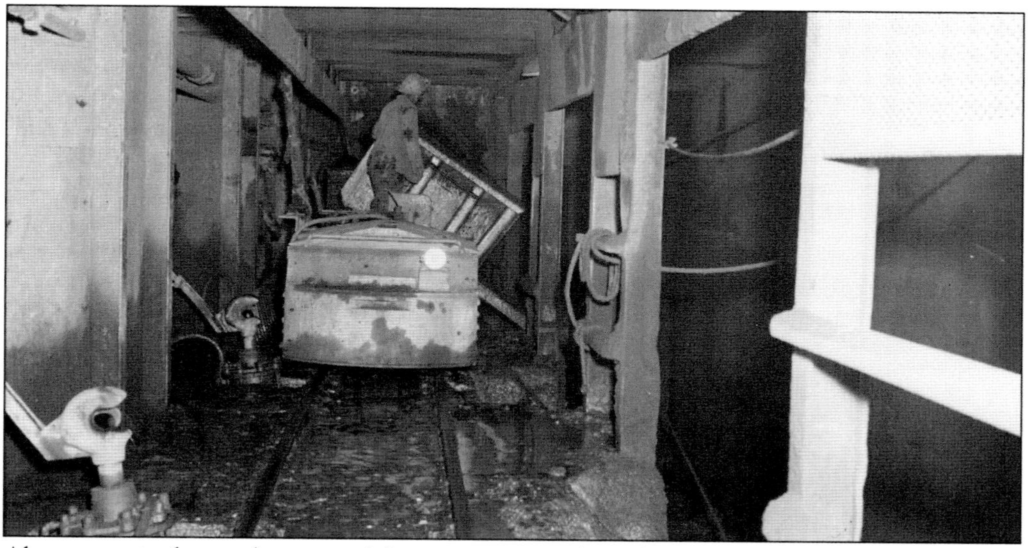

Above, a train dumps the material from a car into a slot. The train would load material from a shaft known as a slushing drift and transport it to a slot. Slushing drifts included an opening to the ore every 20 feet. Miners operated scrapers to move the ore and rock from these openings to areas where the material could fall into the train cars. Trains consisted of six cars, each of which could hold approximately five tons of material. Compressed air cylinders were used to empty the cars. When the material was dumped into the slot, it would fall to a lower level and eventually come out the opening at the bottom of the slot, as shown in the picture below.

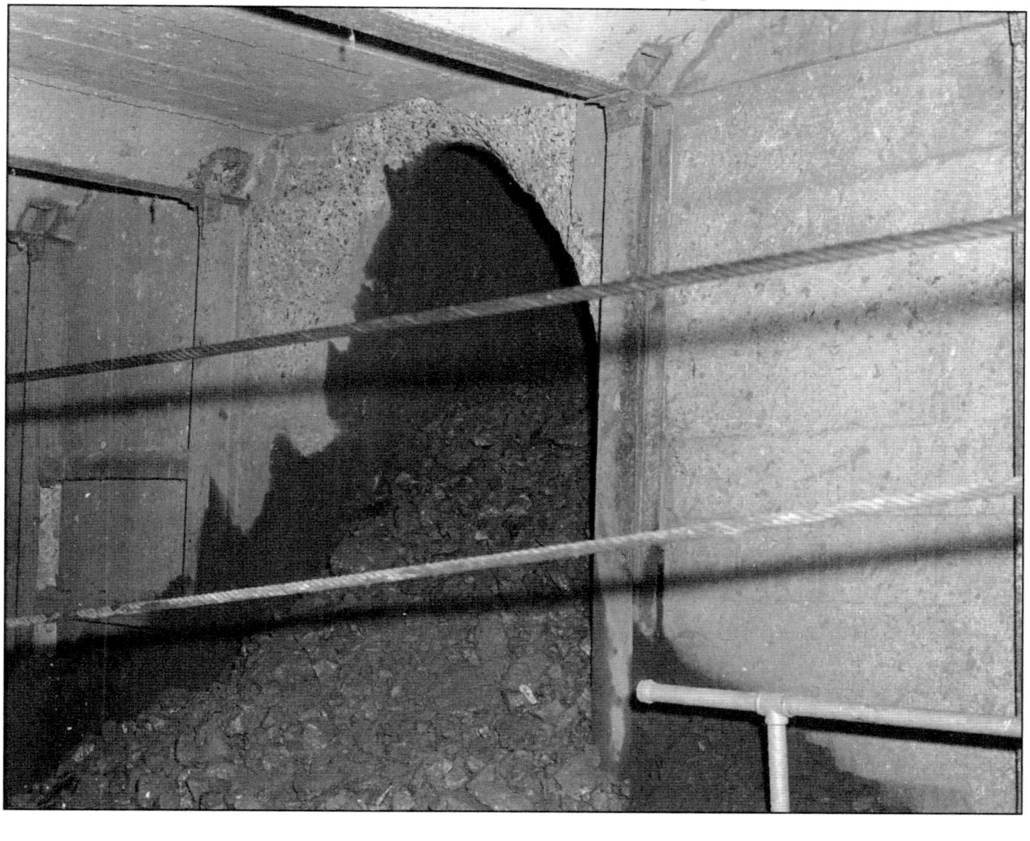

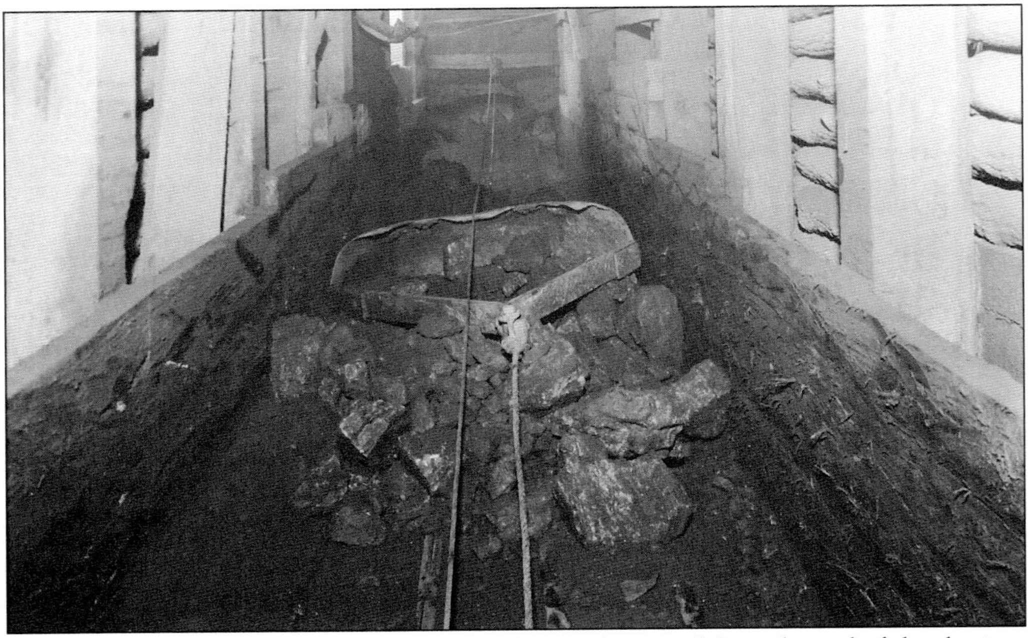

The scraper pictured in the photograph (above) moved material from the end of the slot into openings for a measuring bin. The scraper was powered by an electric motor. The openings to a measuring bin can be seen a short distance in front of the operator (below). When the measuring bin was full, there was enough material to fill the skip. When the skip stopped, the bin was opened, and the material would fill the skip. It was then pulled out of the mine.

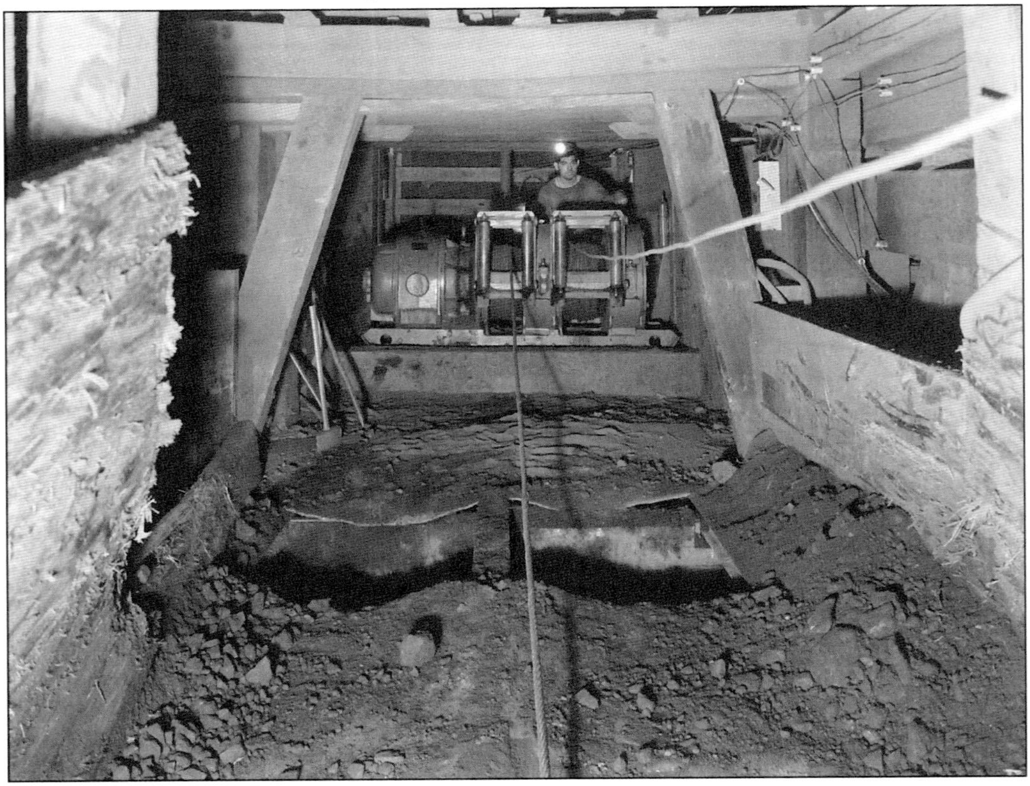

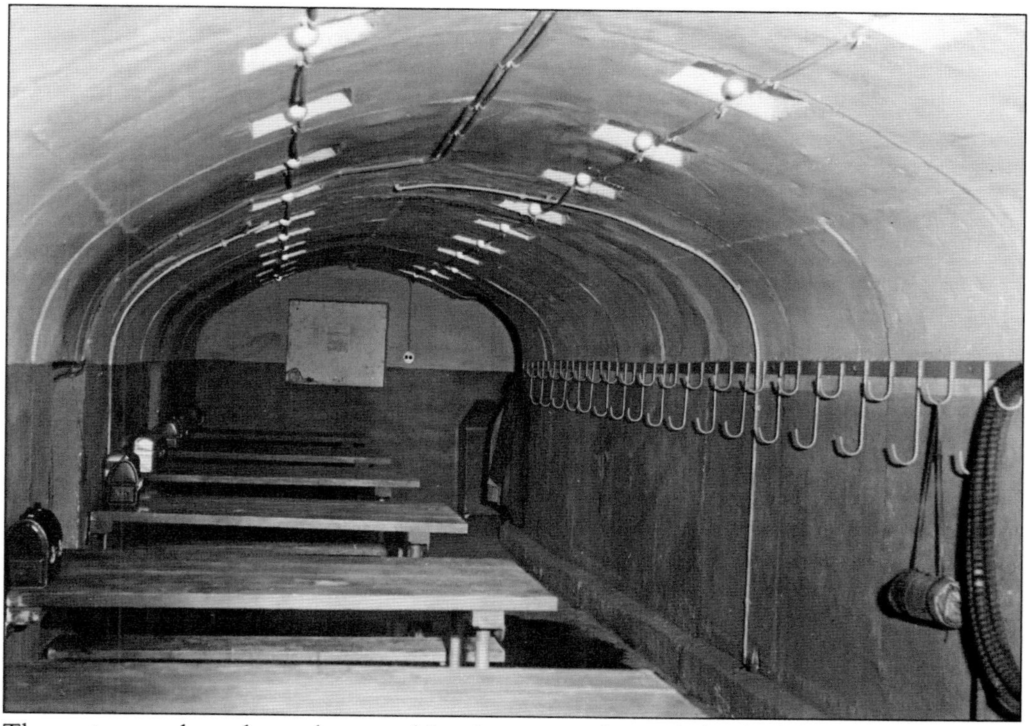

These pictures show the underground lunchroom on the sixth level of the No. 3 Mine. The lunchroom at the No. 4 Mine was not exactly the same, but it was somewhat similar.

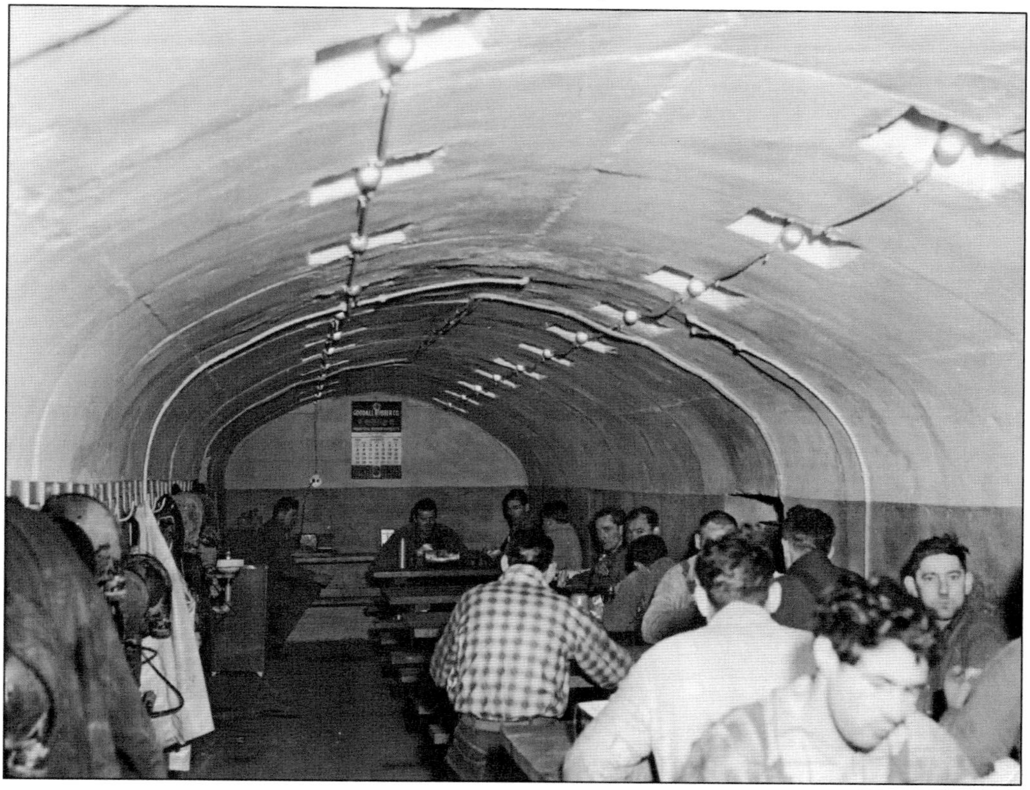

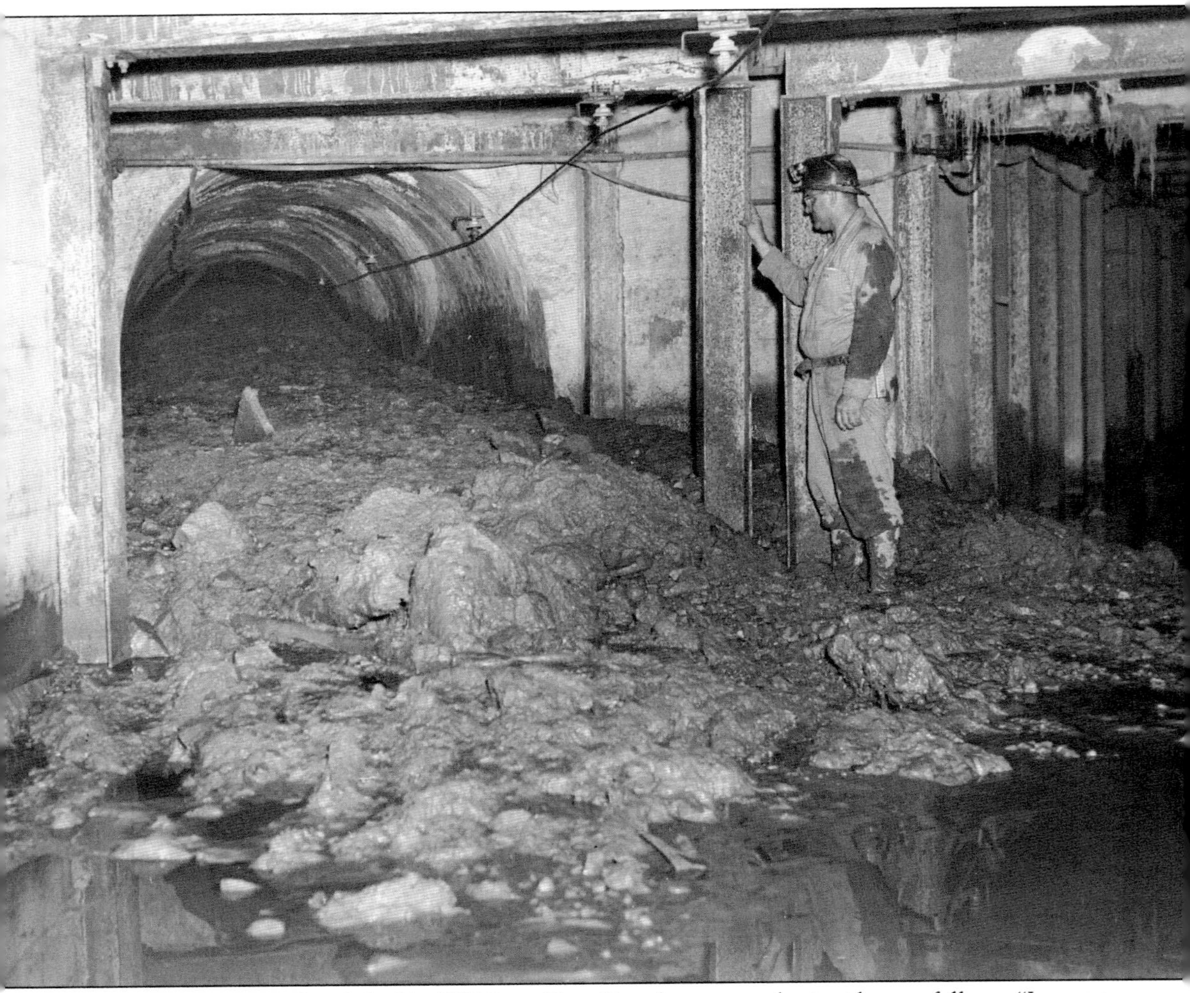

A cave-in occurred on July 24, 1957. Rev. Gary Beard recounts the incident as follows: "I went to work in No. 4 Mine on July 24 with my German textbook in my boot. My job was to operate the train and haul ore gathered from the openings and transport and dump it into the slot. I worked only the afternoon or swing shift since I was a ministerial student at Elizabethtown College. On this particular night I had planned to study German for a German test the next day. Upon arriving at the mine, I found out that Mel Fidler had brought potatoes for a lunchtime treat. The potatoes were placed on the lunchroom heater to cook. I knew it would be a packed noisy lunchroom. The policy of the mine was to park the train at the pocket during lunch but on this night after I dumped the last load I grabbed a potato and some butter then took the train as far from the noisy lunchroom as I could. Using my lamp, I was able to study in peace for about 45 minutes. When I tried to turn the train back on to go back to work, it would not operate. I was scared because I had taken the train away from the pocket, so I had started to walk back to tell some one what I had done. On the way back, I met a supervisor. He asked me what I was doing—I had been listed as missing. I was the only one of the 14 miners not in the lunchroom when this cave-in occurred. Water that was above the ore pushed the ore, mud, and rock through the pocket and into the drift."

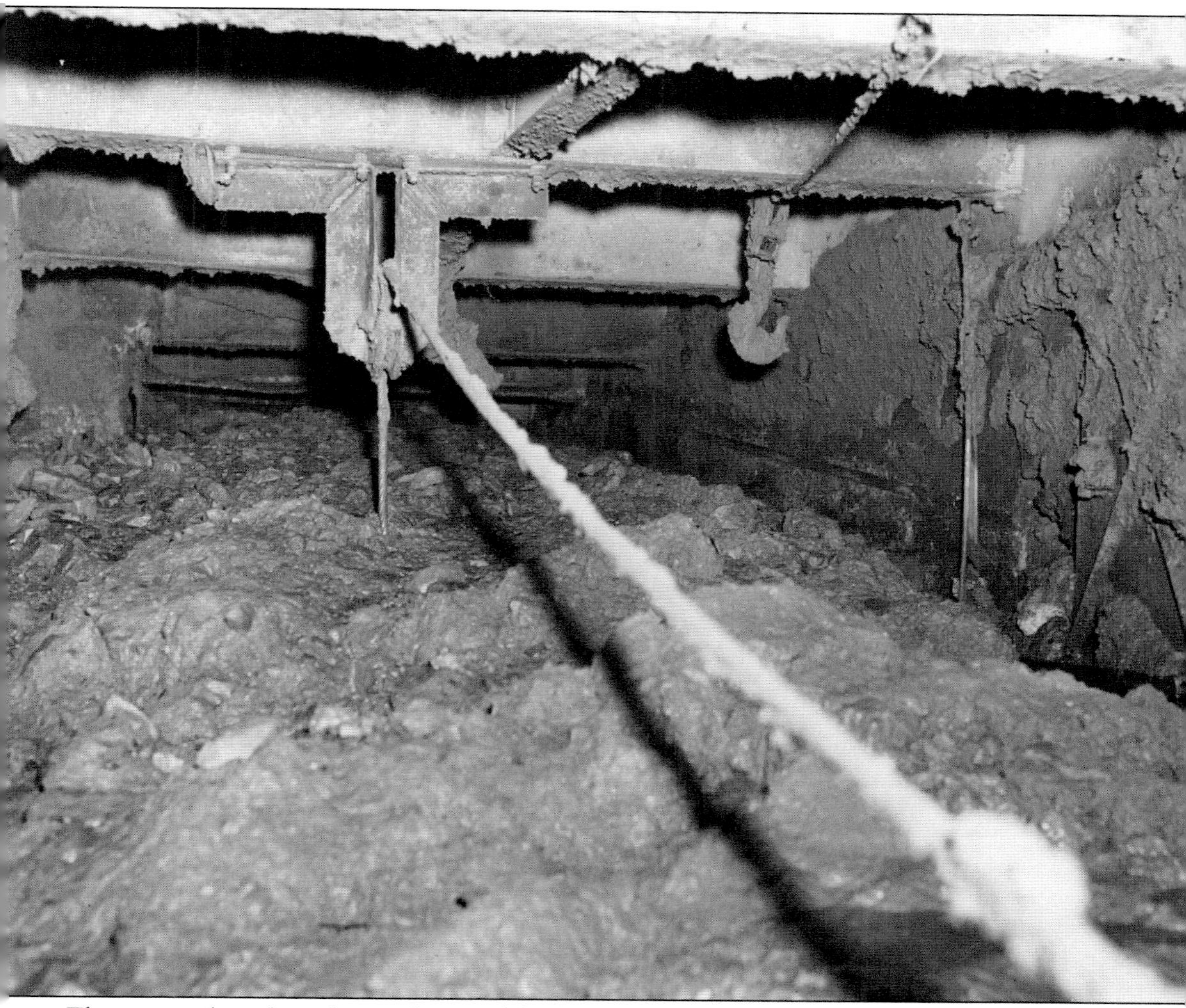

This picture shows how close to the top of the drift debris from the cave-in pushed. There is really nowhere to run to get away from such an incident. Amazingly, no one was injured in the cave-in. Reverend Beard relates what happened after the cave-in: "We finished work and were waiting for the skip to take us up to the outside. While we were waiting, I started to chisel a cross in the bare rock of the loading area. It was small at the beginning but a cross is something that can be extended. Before long, I noticed it was also being worked on by other men on other shifts who were also waiting for the skip to take them out of the mine after their shift was completed. Soon, it was a beautiful impression and expression of faith especially after the cave-in. It was the first thing I looked at when we started work and the last impression I saw at the end of the day."

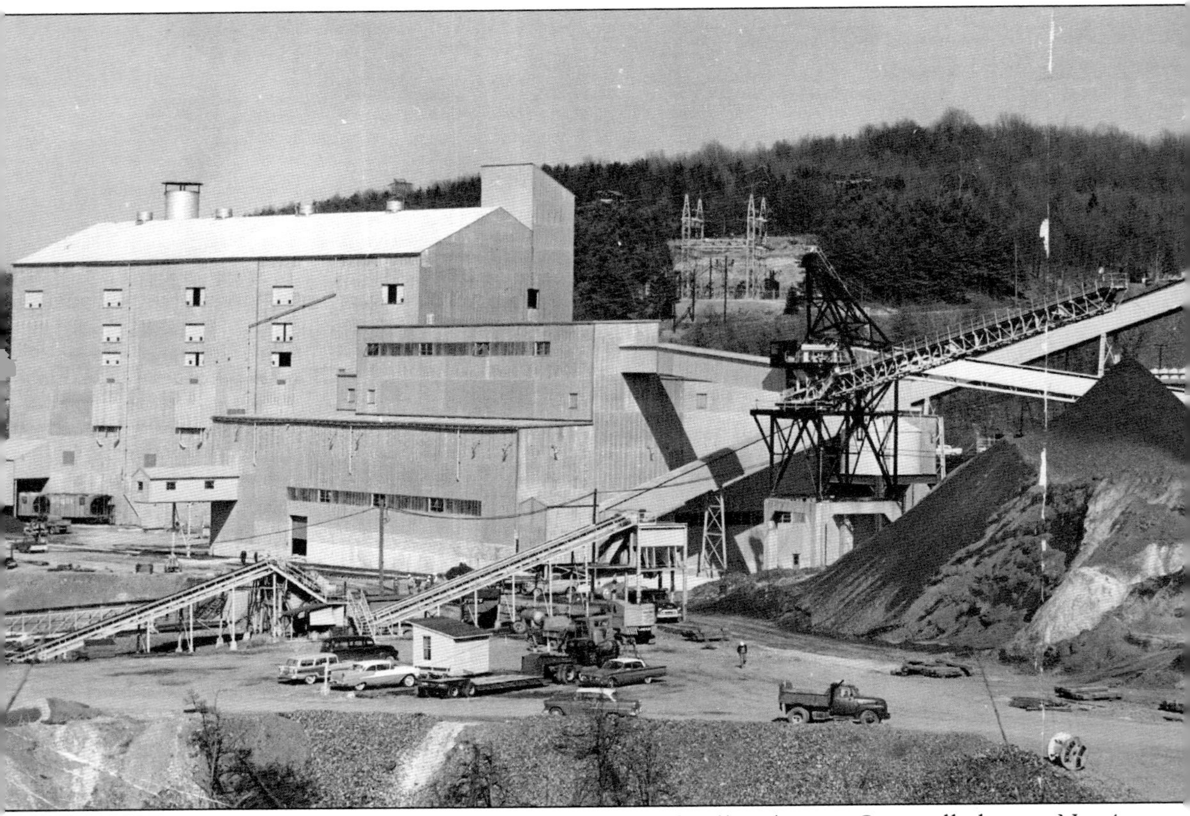

In 1962, Bethlehem Steel opened a modern mill and pellet plant in Cornwall close to No. 4 Mine. At the same time, the Lebanon Concentrator Plant was closed and torn down after 57 years of operation. The pellet plant crushed the ore and formed it into round pellets. These were energy efficient in the blast furnaces because heat and air could be circulated around them. This photograph shows the plant and the conveyor system used to supply it with ore.

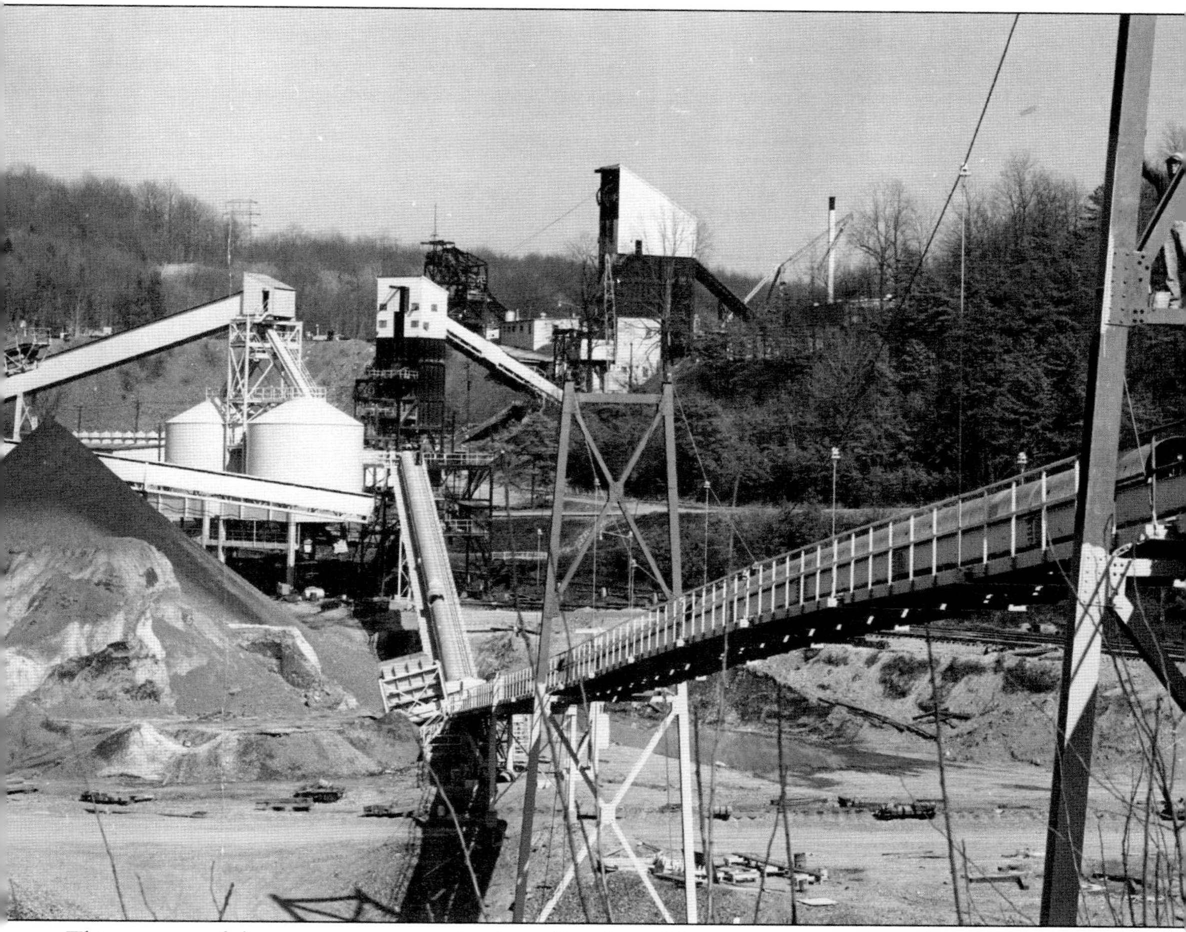

This is more of the conveyor system for the pellet plant. The section going from the right side to about the center is the end of the mile-long conveyor system that brought ore from No. 3 Mine.

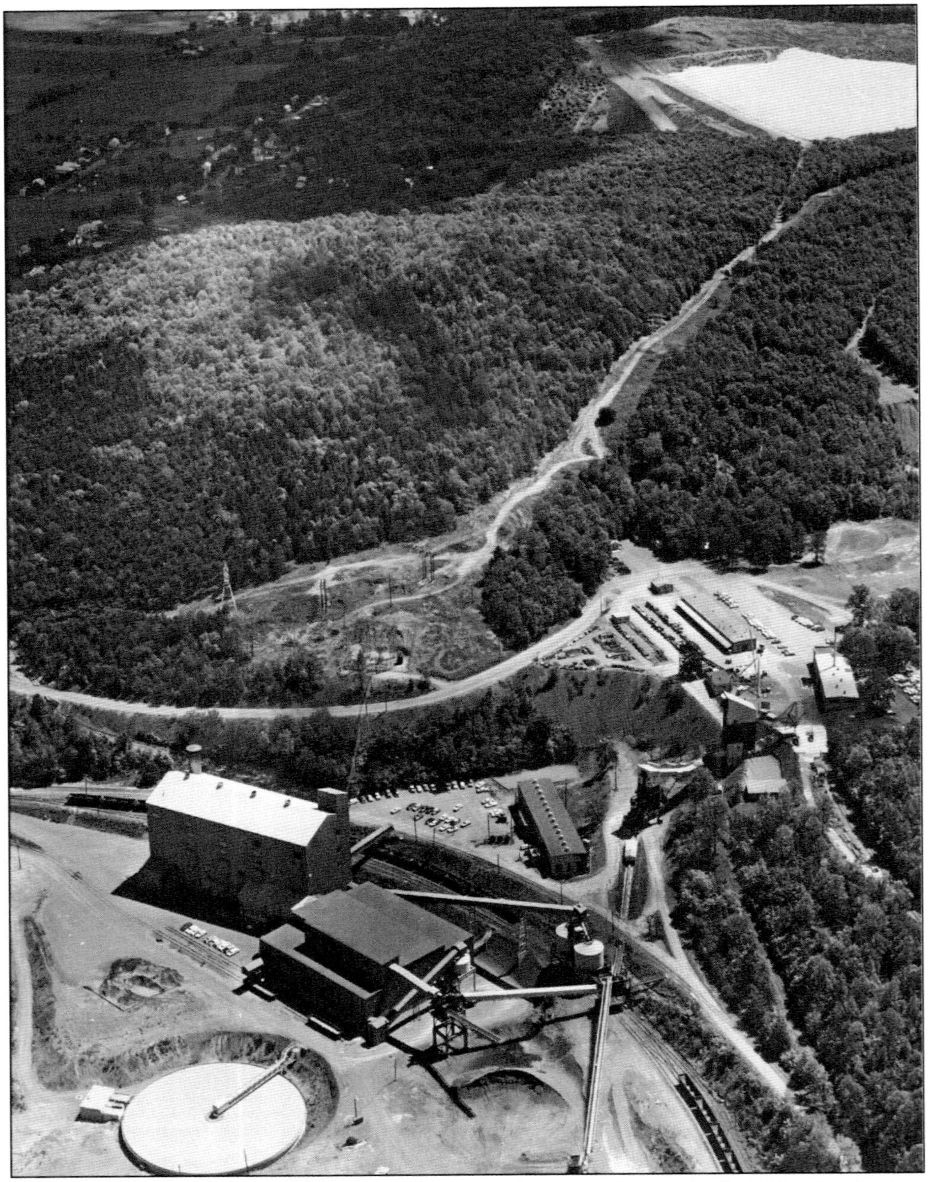

The bottom portion of this aerial photograph shows the pellet plant. The building in the forefront houses a milling process, in which ore was milled to a fine powder. Ore slurry would be pumped to either the taller building on the left or the round overage holding tank. Once inside the taller structure, the slurry was made into pellets, which were shipped by railcar to Bethlehem Steel's blast furnaces. To the right and below the plant is the hoist house, from which the skips were controlled. At the top left of the cluster of structures is a maintenance facility, where mechanics and electricians had shops. It also contained a small machine shop. The next structure, below and to the right with the slanted roof, is the tipple, or headframe, of No. 4 Mine. Skips loaded with ore were pulled up the incline by cable from the hoist house and continued until they were tipped and dumped ore into the tipple. The ore would be crushed, separated from waste rock, and sent by conveyor belt to the pellet plant. The building by the trees on the far right is the office and dry house, the area where the miners changed clothes. At the top right of the image is a settling dam. Water containing pellet process tailings would be pumped there for particulates to settle.

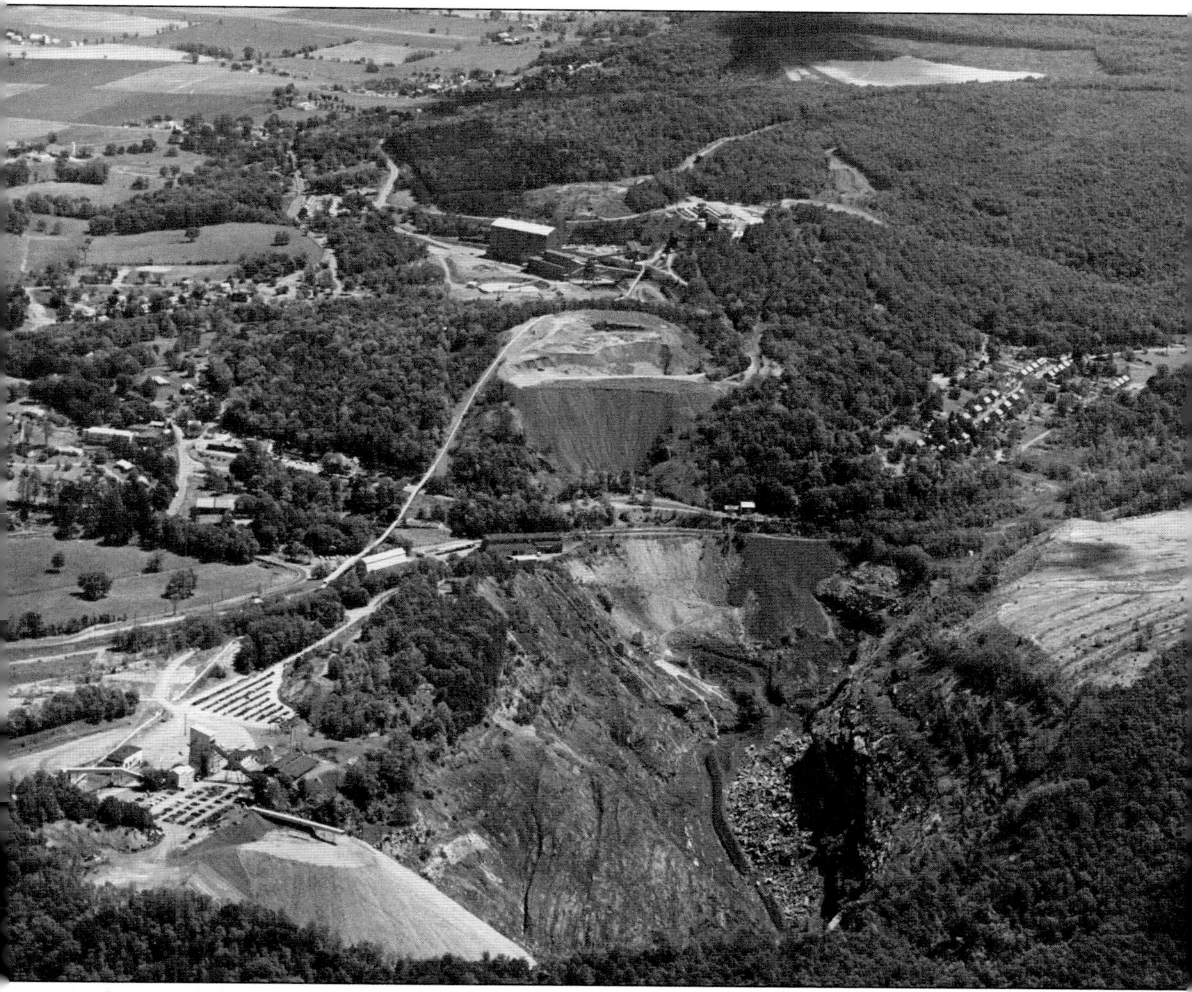

In this aerial photograph, the No. 3 Mine is in the lower left corner. The open pit is empty. This is well after the work in the pit ended. The No. 3 Mine has a conveyor that is taking waste rock and putting it back into the pit rather than trucking it elsewhere. From the crusher, a short conveyor takes the ore to the mile-long system that ends at the pellet plant. The system appears to be a white line extending from the No. 3 Mine to the pellet plant. On the right at about the middle of the picture is Minersvillage. To the left of the conveyor is the Cornwall Iron Furnace. At the left edge are the buildings of Cornwall Manor. The little village that appears above Cornwall Manor is Anthracite.

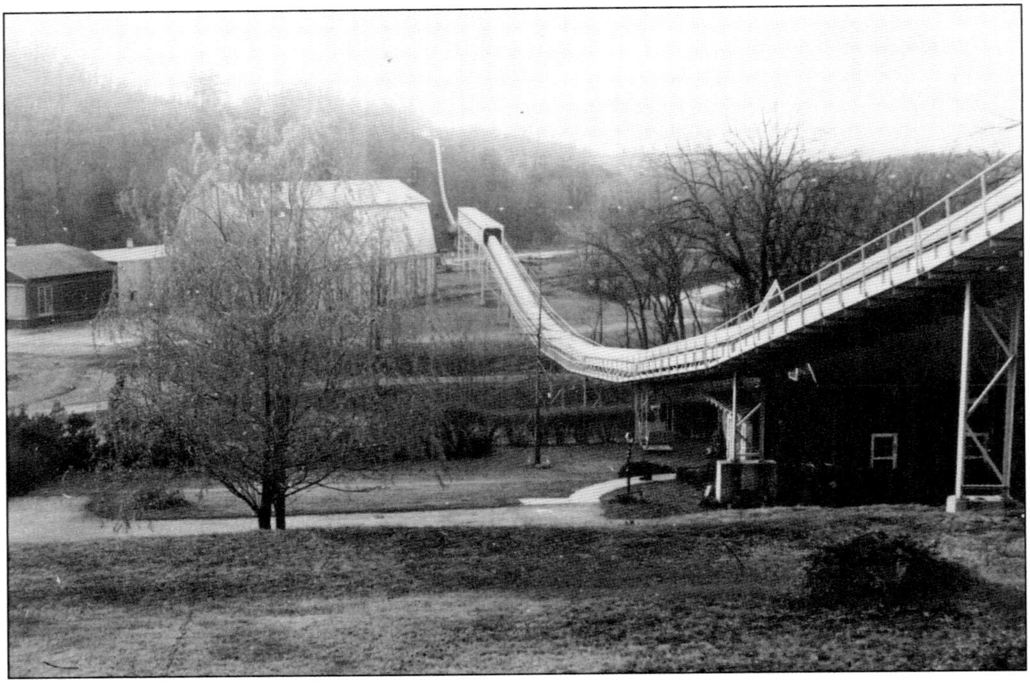

The picture above shows the elevation of the conveyor system from No. 3 Mine past the loading pocket as it climbs up Big Hill on its way to the pellet plant. Below are some of the actual pellets produced at Cornwall. The sizes vary, and they are not all exactly round; nonetheless, these pellets saved energy when they were used in a blast furnace.

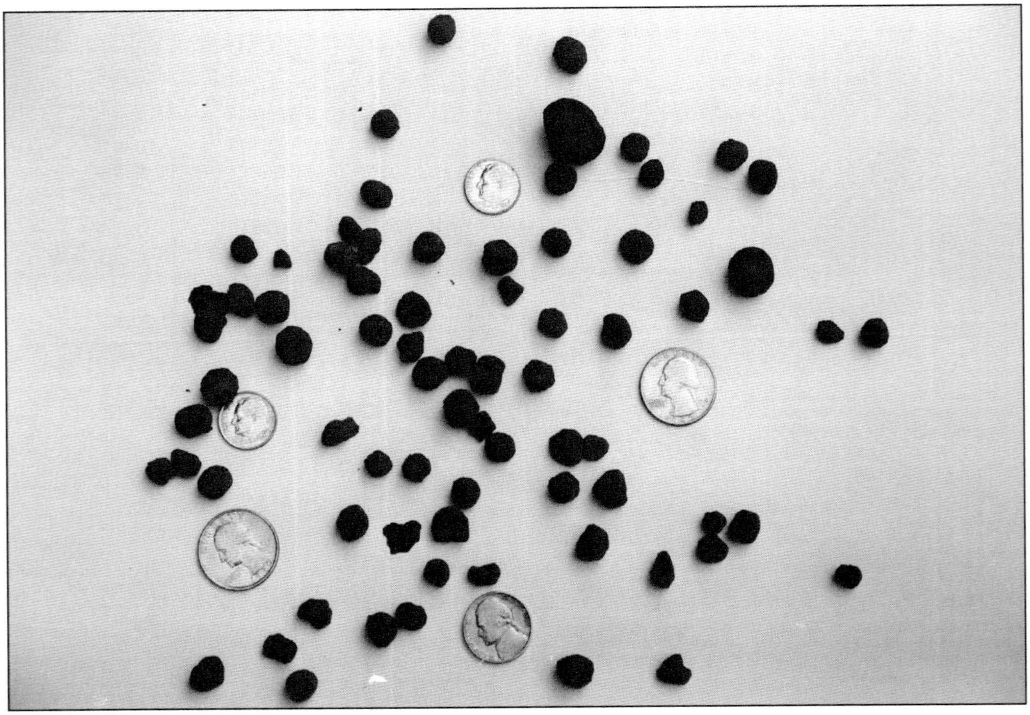

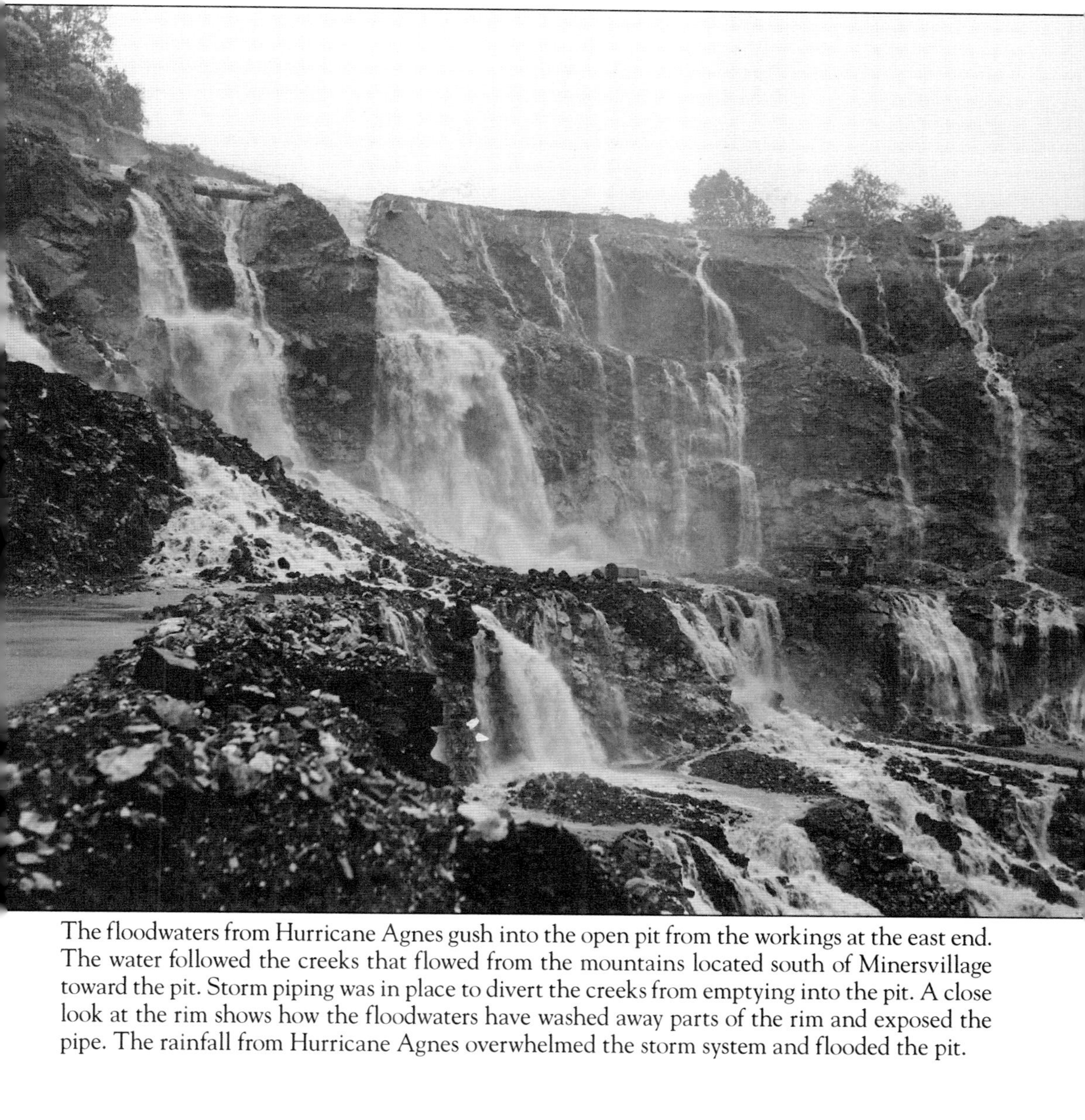

The floodwaters from Hurricane Agnes gush into the open pit from the workings at the east end. The water followed the creeks that flowed from the mountains located south of Minersvillage toward the pit. Storm piping was in place to divert the creeks from emptying into the pit. A close look at the rim shows how the floodwaters have washed away parts of the rim and exposed the pipe. The rainfall from Hurricane Agnes overwhelmed the storm system and flooded the pit.

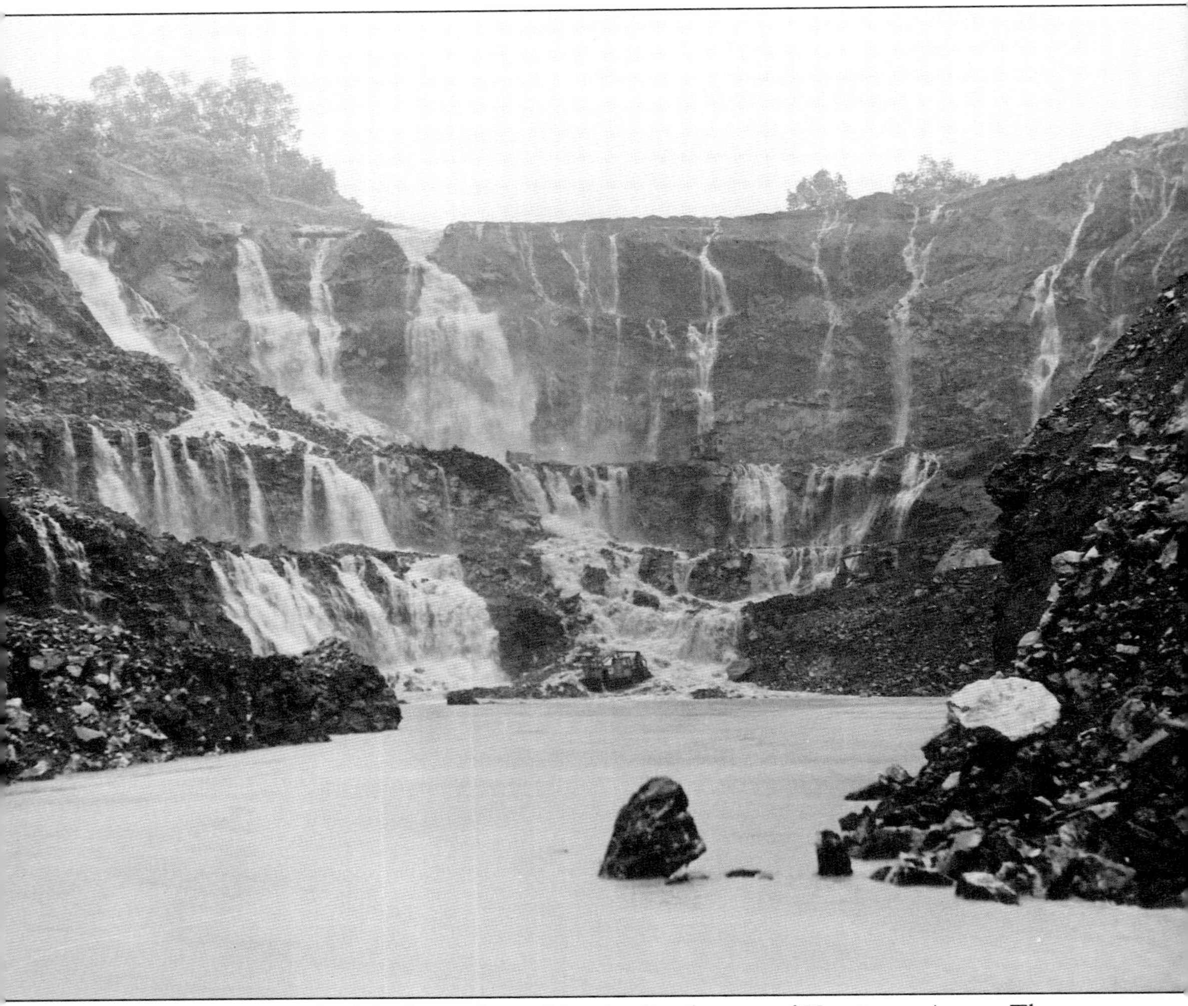

This image shows the lake being formed by the floodwaters of Hurricane Agnes. The water entered No. 3 Mine from openings where the mine had broken through into the open pit and through the dumping pockets. The underground workings of No. 3 Mine were totally flooded after electricity was lost, and the pumps were inundated with water.

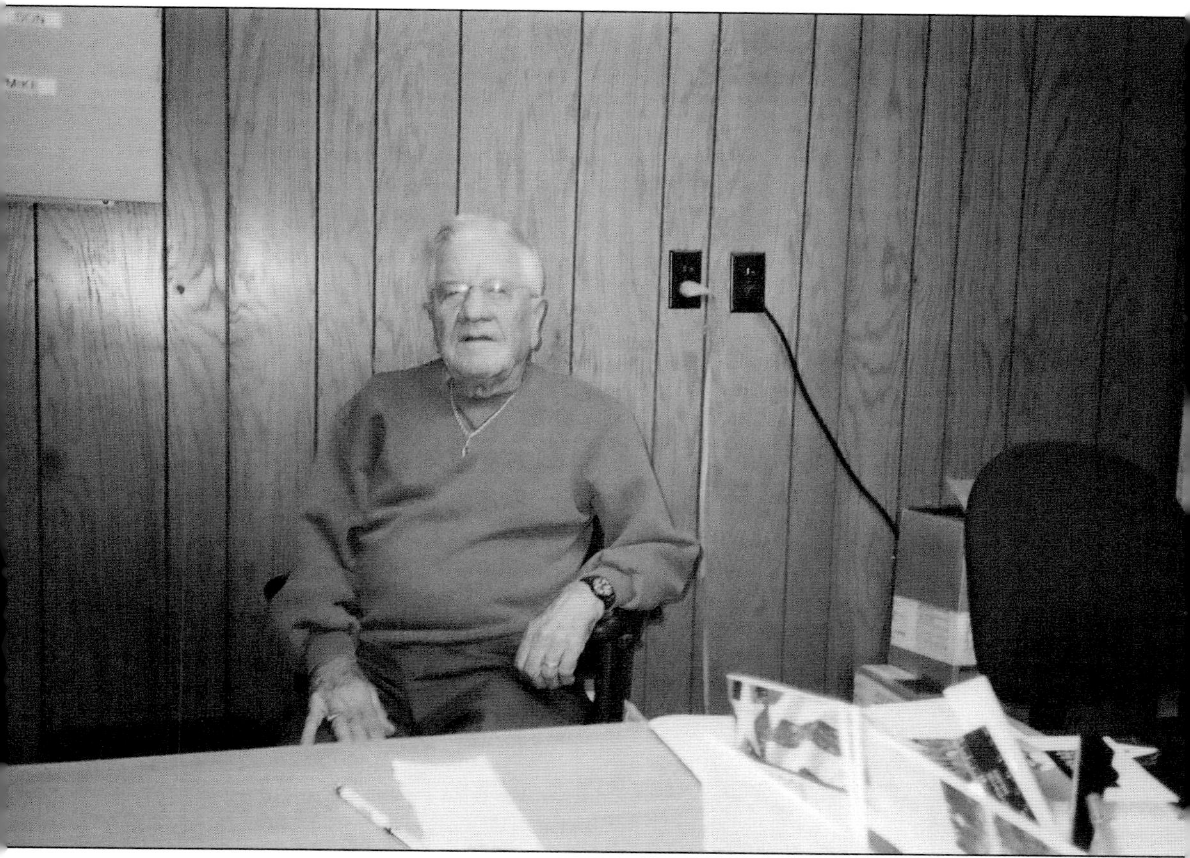

Milan Lepinsky was the last mine superintendent at the Bethlehem Steel Mines at Cornwall. When interviewed, Lepinsky explained that the mines were using a rather odd electrical system. Electricity for the mines was transmitted to Cornwall from the power station at the Bethlehem Steel plant in Lebanon. The power station converted the electricity from the standard 60-cycle power that was generated by the electric company and used in private homes to 25-cycle power that was used in the underground mines. Both No. 3 and No. 4 Mines and the supporting equipment used 25-cycle power. On the day of the flood, the water level rose very quickly at the power plant in Lebanon, and the floodwater ruined the machinery that converted the electricity to a usable form, leaving the mines without power. Soon thereafter, the Bethlehem Steel corporate office decided to close the mines even though ore was still present. The cost of reopening the mines was estimated to exceed the value of the ore.

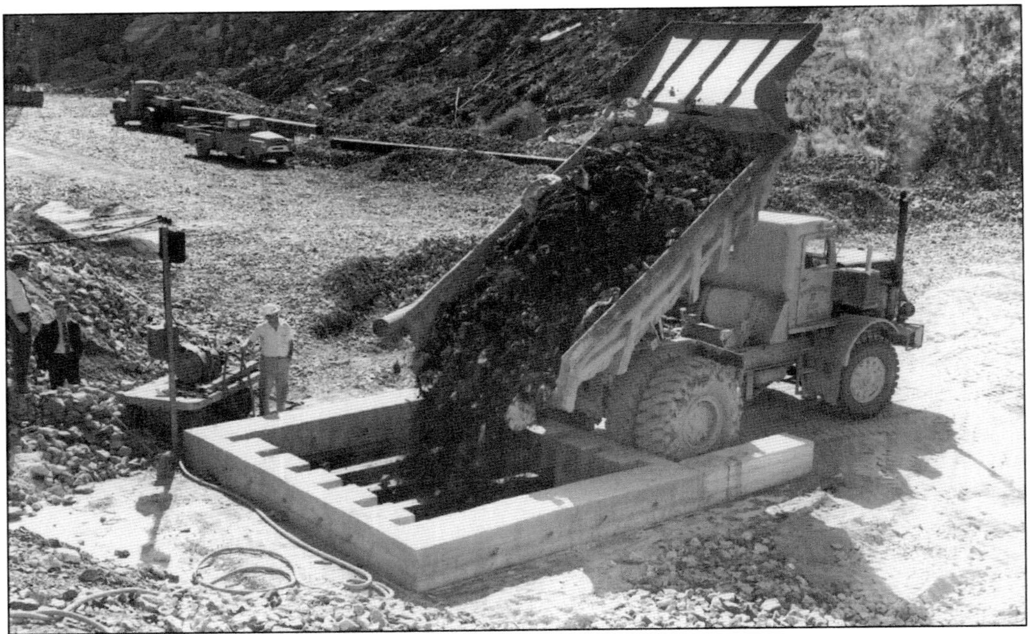

The open pit was expanded eastward in the latter part of the 1960s to mine the east-end ore body. Above, a 1970 image shows the dumping pocket that would allow the ore to fall to the No. 3 Mine underground. The ore could then be taken out via the skips and processed. Below, the very last truckload of ore mined from the east-end ore body leaves the mine. This was on June 30, 1973, about one year after Hurricane Agnes flooded the underground No. 3 Mine and began filling the large open pit with water. Because the mine was waterlogged, the truck is hauling the ore to the crusher rather than dumping it into the No. 3 Mine dumping pocket. With this load, the mining operation was closed forever at Cornwall. (Below, courtesy of Daniel Chernich.)

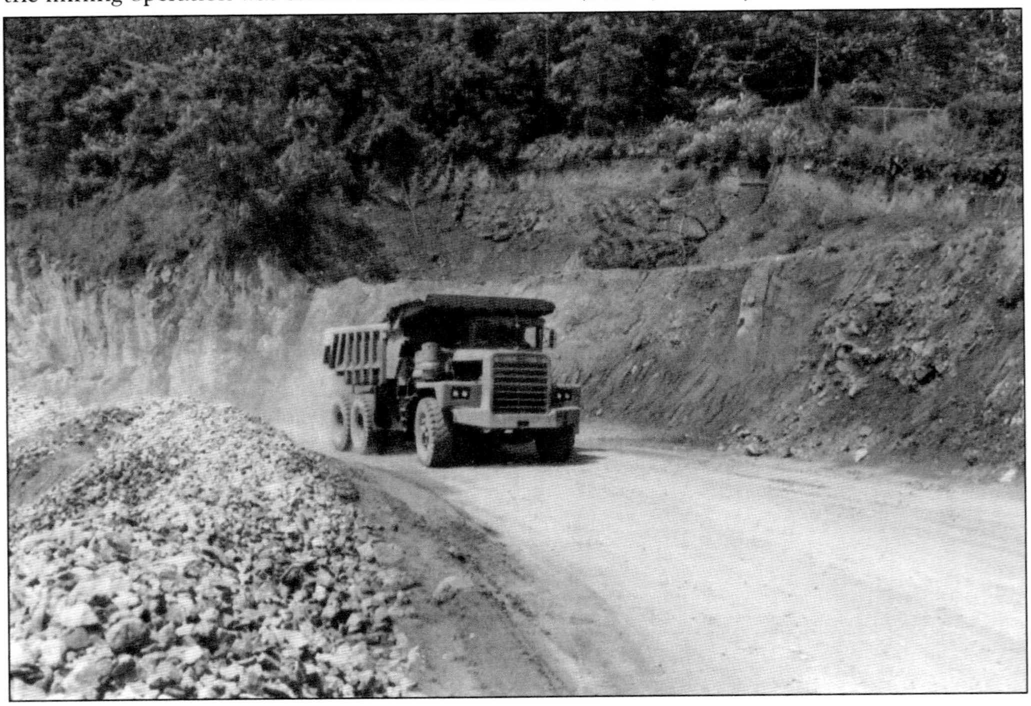

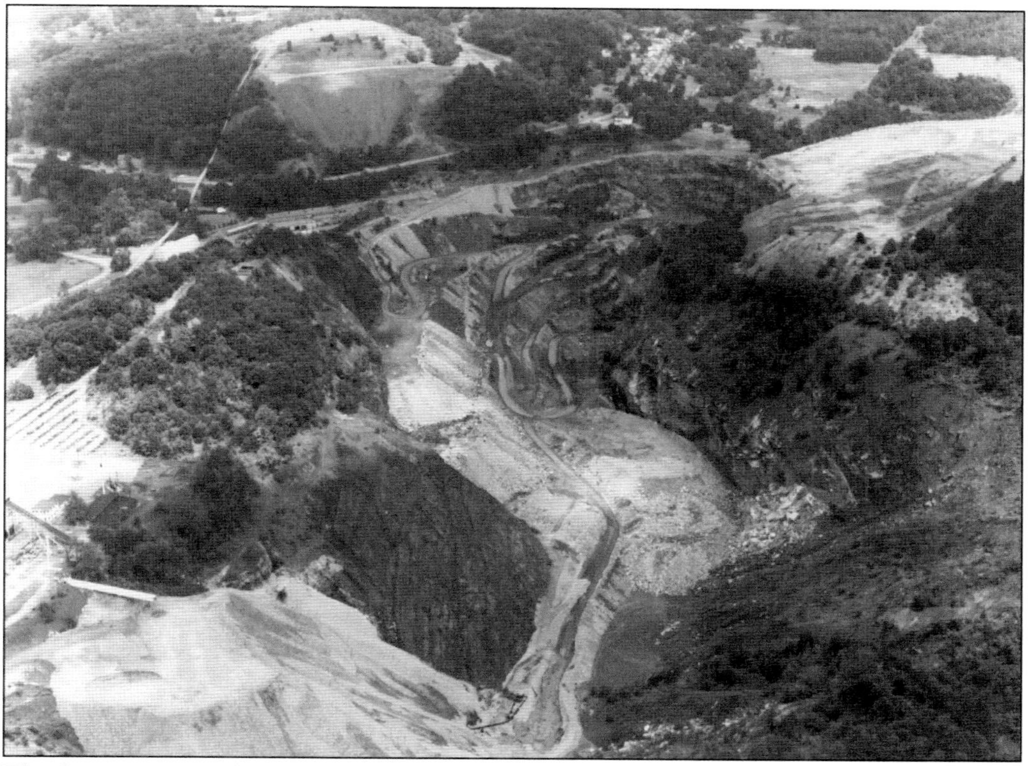

This June 29, 1973, photograph (above) shows the open pit on the next-to-last day of mining in the east-end ore body a year after Hurricane Agnes flooded No. 3 Mine and caused a loss of electricity at both mines. The mines may have been able to recover from the flood had the power not been lost. A recent image (below) shows the open pit as it looks today. Nature, trees, and water have obscured almost all evidence of what was a very busy mining operation. At one time, open-pit miners, underground miners, administrative employees, and pellet plant employees—a total of 900 people—earned their living at this site.

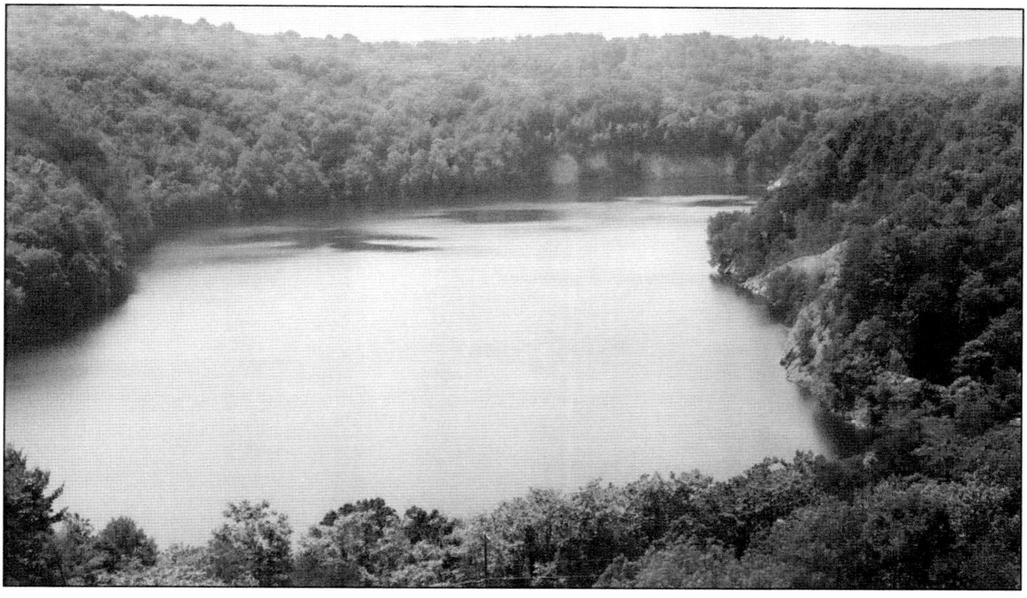

F

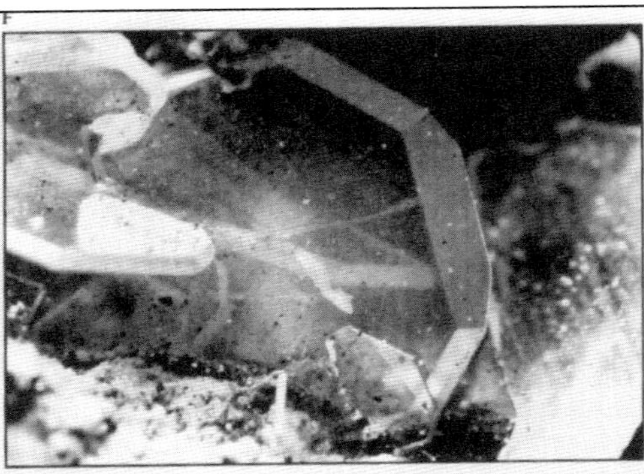

G

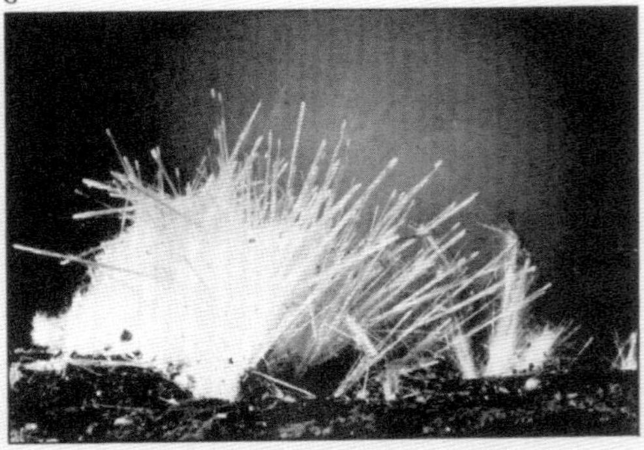

H

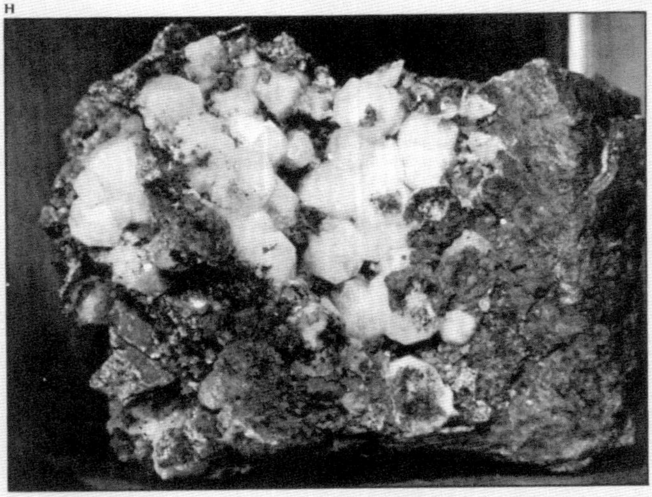

I

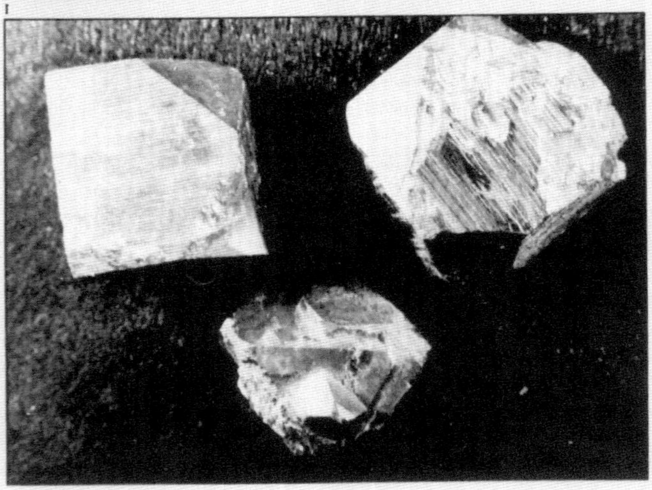

CURRENT CORNWALL SPECIES LIST

Actinolite
(byssolite)
Albite
Analcime
Andradite
Antigorite
Apatite group
Apophyllite
Aragonite
Augite
Biotite
Brochantite (?)
Brucite
Calcite
Chalcanthite
Chalcocite
Chalcopyrite
Chlorite group
Chrysotile
Chrysocolla
Clinozoisite
Copper
Cuprite
Datolite
Diopside
Elbaite (?)
Enstatite
Epidote
Erythrite
Fluorite
Forsterite
Galena
Grossular
Gypsum
Hematite
Heulandite
Hornblende
Hypersthene
Ilmenite
Magnetite
Malachite
Microcline
Millerite
Muscovite
Natrolite
Orthoclase
Phlogopite
Pigeonite
Prehnite
Pyrite
Quartz
Rutile (?)
Stilbite
Sphalerite
Talc
Titanite
Tochilinite
Tremolite
Vesuvianite
Wurtzite

After the mines closed, Cornwall became famous for the varied assortment of minerals that could be found at the rock dumpsites. Mineral societies and collectors from Pennsylvania and neighboring states have visited in search of finds. This image includes a complete listing of minerals found in Cornwall. Some of these were discovered only in small quantities and have not turned up for many years. Entries marked with a question mark have not been verified.

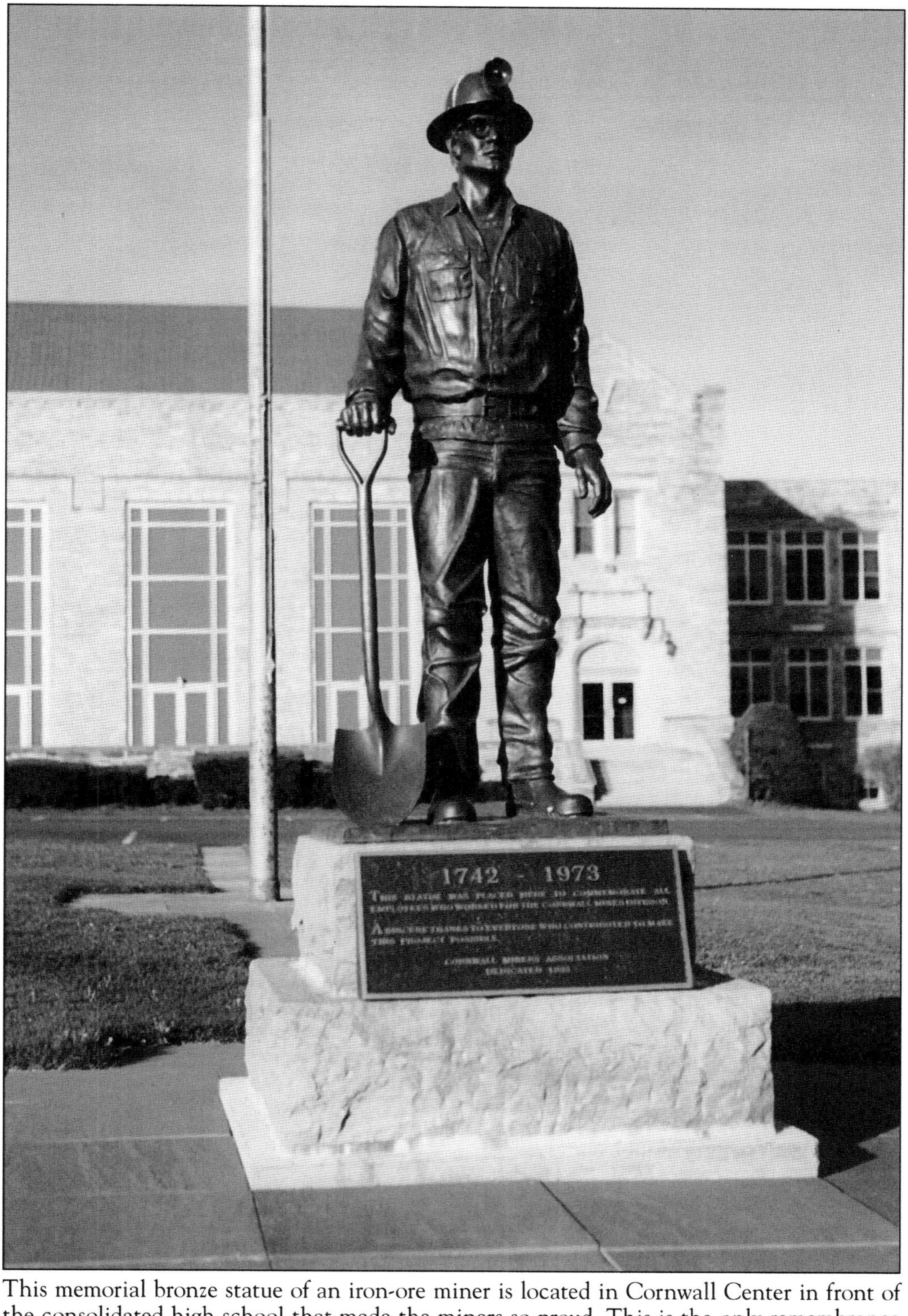

This memorial bronze statue of an iron-ore miner is located in Cornwall Center in front of the consolidated high school that made the miners so proud. This is the only remembrance of the men who did the hard work of mining iron ore for the 236 years (1737–1973) that the mines operated.